First Principles
Part II Laws Of The Knowable

by

Herbert Spencer

First Principles
Part II Laws Of The Knowable
by Herbert Spencer

Copyright © 2024

All Rights reserved.

ISBN: 978-93-61429-29-3

Published by

DOUBLE 9 BOOKS

2/13-B, Ansari Road
Daryaganj, New Delhi – 110002
info@double9books.com
www.double9books.com
Tel. 011-40042856

This book is under public domain

ABOUT THE AUTHOR

English polymath Herbert Spencer worked as a sociologist, anthropological, biologist, psychologist, and philosopher. The phrase "survival of the fittest" was first used by Spencer in Principles of Biology (1864), following his reading of Charles Darwin's 1859 book On the Origin of Species. Although the name primarily denotes natural selection, Spencer also embraced Lamarckism since he believed that evolution extends into the fields of sociology and ethics. Spencer created a comprehensive theory of evolution that included the progressive development of biological systems, the physical environment, human thought, culture, and society. He made contributions to many different fields as a polymath, such as politics, economics, anthropology, ethics, literature, astronomy, biology, sociology, and psychology. He attained great power throughout his lifetime, mostly in academic English-speaking circles. Although Spencer was "the single most famous European intellectual in the closing decades of the nineteenth century," his impact began to wane after 1900. Talcott Parsons questioned, "Who now reads Spencer?" in 1937. Spencer, the son of William George Spencer (often referred to as George), was born in Derby, England, on April 27, 1820.

CONTENTS

CHAPTER I
LAWS IN GENERAL

§ 35. We have seen that intellectual advance has been dual—has been towards the establishment of both a positively unknown and a positively known. In making ever more certain the inaccessibility of one kind of truth, experience has made ever more certain the accessibility of another kind. The differentiation of the knowable from the unknowable, is shown as much in the reduction of the one to perfect clearness, as in the reduction of the other to impenetrable mystery. Progressing enlightenment discloses a definite limit to human intelligence; and while all which lies on the other side of the limit, is, with increasing distinctness, seen to transcend our finite faculties, it grows more and more obvious that all which lies on this side of the limit may become an indisputable possession.

To speak specifically—it has been shown that though we can never learn the *nature* of that which is manifested to us, we are daily learning more completely the *order* of its manifestations. We are conscious of effects produced in us by something separate from ourselves. The effects of which we are conscious—the changes of consciousness which make up our mental life, we ascribe to the forces of an external world. The intrinsic character of these forces—of this external world—of that which underlies all appearances, we find inscrutable; as is also the internal something whose changes constitute consciousness, but at the same time we find that among the changes of consciousness thus produced, there exist various constant relations; and we have no choice but to ascribe constancy to the relations which subsist among the inscrutable causes of these changes. Observation early discloses certain invariable connexions of coexistence and sequence among phenomena. Accumulating experiences tend continually to augment the number of invariable connexions recognized. When, as in the later stages of civilization, there arises not only a diligent gathering together of experiences but a critical comparison of them, more remote and complex connexions are added to the list. And gradually there grows up the habit of regarding these uniformities of relation as characterizing all manifestations of the Unknowable. Under the endless variety and seeming irregularity, there is ever more clearly discerned that "constant course of procedure" which we call Law.

The growing belief in the universality of Law, is so conspicuous to all cultivated minds as scarcely to need illustration. None who read these pages will ask for proof that this has been the central element of intellectual progress. But though the fact is sufficiently familiar, the philosophy of the fact is not so; and it will be desirable now to consider it. Partly because the development of our conception of Law will so be rendered more comprehensible; but chiefly because our subsequent course will thus be facilitated; I propose here to enumerate the several conditions that determine the order in which the various relations among phenomena are discovered. Seeing, as we shall, the consequent necessity of this order; and enabled, as we shall also be, to estimate the future by inference from the past; we shall perceive how inevitable is our advance towards the ultimatum that has been indicated.

§ 36. The recognition of Law, being the recognition of uniformity of relations among phenomena, it follows that the order in which different groups of phenomena are reduced to law, must depend on the frequency and distinctness with which the uniform relations they severally present, are experienced. At any given stage of progress, those uniformities will be most recognized with which men's minds have been oftenest and most strongly impressed. In proportion partly to the number of times a relation has been presented to consciousness (not merely to the senses); and in proportion partly to the vividness with which the terms of the relation have been cognized; will be the degree in which the constancy of connexion is perceived.

The frequency and impressiveness with which different classes of relations are repeated in conscious experience, thus primarily determining the succession in which they are generalized, there result certain derivative principles to which this succession must more immediately and obviously conform. First in importance comes *the directness with which personal welfare is affected*. While, among surrounding things, many do not appreciably influence the body in any way, some act detrimentally and some beneficially, in various degrees; and manifestly, those things whose actions on the organism are most influential, will, cæteris paribus, be those whose laws of action are earliest observed. Second in order, is *the conspicuousness of one or both the phenomena between which a relation is to be perceived*. On every side are countless phenomena so concealed as to be detected only by close observation; others not obtrusive enough to attract notice; others which moderately solicit the attention; others so imposing or vivid as to force themselves upon consciousness; and supposing incidental conditions to be the same, these last will of course be among the first to have their relations generalized. In the third place, we have *the absolute frequency with which the relations occur*. There are coexistences and sequences of all degrees of commonness, from those which are ever present to those which are extremely rare; and it is clear that the rare coexistences and sequences, as well as the sequences which are very long in taking place, will not be reduced to law so soon as those which are familiar and rapid. Fourthly has to be added *the relative frequency of occurrence*. Many events and appearances are more or less limited to times and places; and as a relation which does not exist within the environment of an observer, cannot be cognized by him, however common it may be elsewhere or in another age, we have to take account of the surrounding physical circumstances, as well as the state of society, of the arts, and of the sciences—all of which affect the frequency with which certain groups of facts are exposed to observation. The fifth corollary to be noticed, is, that the succession in which different classes of phenomena are reduced to law, depends in part on their *simplicity*. Phenomena presenting great composition of causes or conditions, have their essential relations so masked, that it requires accumulated experiences to impress upon consciousness the true connexion of antecedents and consequents they involve. Hence, other things equal, the progress of generalization will be from the simple to the complex; and this it is which M. Comte has wrongly asserted to be the sole regulative principle of the progress. Sixth, and last, comes *the degree of abstractness*. Concrete relations are the earliest acquisitions. The colligation of any group of these into a general relation, which is the first step in abstraction, necessarily comes later than the discovery of the relations colligated. The union of a number of these lowest generalizations into a higher and more abstract generalization, is necessarily subsequent to the formation of such lowest generalizations. And so on continually, until the highest and most abstract generalizations have been reached.

These then are the several derivative principles. The frequency and vividness with which uniform relations are repeated in conscious experience, determining the recognition of their uniformity; and this frequency and vividness depending on the above conditions; it follows that the order in which different classes of facts are generalized, must depend on the extent to which the above conditions are fulfilled in each class. Let us mark how the facts harmonize with this conclusion: taking first a few that elucidate the general truth, and afterwards some that are illustrative of the several special truths which we here see follow from it.

§ 37. The relations earliest known as uniformities, are those subsisting between the common physical properties of matter—tangibility, visibility, cohesion, weight &c. We have no trace of an era in human history when the resistance offered by every visible object, was regarded as caused by the will of the object; or when the pressure of a body on the hand supporting it, was ascribed to the direct agency of a living being. And accordingly, we see that these are the relations oftenest repeated in consciousness; being as they are, objectively frequent, conspicuous, simple, concrete, and of immediate personal concern.

Similarly with respect to the ordinary phenomena of motion. The fall of a mass on the withdrawal of its support, is a sequence which directly affects bodily welfare, is conspicuous, simple, concrete, and very often repeated. Hence it is one of the uniformities recognized before the dawn of tradition. We know of no time when movements due to terrestrial gravitation were attributed to volition. Only when the relation is obscured—only, as in the case of an aerolite, where the antecedent of the descent is unperceived, do we find the fetishistic conception persistent. On the other hand, motions of intrinsically the same order as that of a falling stone—those of the heavenly bodies—long remain ungeneralized; and until their uniformity is seen, are construed as results of will. This difference is clearly not dependent on comparative complexity or abstractness; since the motion of a planet in an ellipse, is as simple and concrete a phenomenon as the motion of a projected arrow in a parabola. But the antecedents are not conspicuous; the sequences are of long duration; and they are infrequently repeated. Hence in a given period, there cannot be the same multiplied experiences of them. And that this is the chief cause of their slow reduction to law, we see in the fact that they are severally generalized in the order of their frequency and conspicuousness—the moon's monthly cycle, the sun's annual change, the periods of the inferior planets, the periods of the superior planets.

While astronomical sequences were still ascribed to volition, certain terrestrial sequences of a different kind, but some of them equally without complication, were interpreted in like manner. The solidification of water at a low temperature, is a phenomenon that is simple, concrete, and of much personal concern. But it is neither so frequent as those which we saw are earliest generalized, nor is the presence of the antecedent so uniformly conspicuous. Though in all but tropical climates, mid-winter displays the relation between cold and freezing with tolerable constancy; yet, during the spring and autumn, the occasional appearance of ice in the mornings has no very manifest connexion with coldness of the weather. Sensation being so inaccurate a measure, it is not possible for the savage to experience the definite relation between a temperature of 32° and the congealing of water; and hence the long-continued conception of personal agency. Similarly, but still more clearly, with the winds. The absence of regularity and the inconspicuousness of the antecedents, allowed the mythological explanation to survive for a great period.

During the era in which the uniformity of many quite simple inorganic relations was still unrecognized, certain classes of organic relations, intrinsically very complex and special, were generalized. The constant coexistence of feathers and a beak, of four legs with a bony internal framework, of a particular leaf with poisonous berries, are facts which were, and are, familiar to every savage. Did a savage find a bird with teeth, or a mammal clothed with feathers, he would be as much surprised as an instructed naturalist; and would probably make a fetish of the anomalous form: so showing that while the exceptional relation suggested the notion of a personal cause, the habitual relation did not. Now these uniformities of organic structure which are so early perceived, are of exactly the same class as those more numerous ones later established by biology. The constant coexistence of mammary glands with two occipital condyles in the skull, of vertebræ with teeth lodged in sockets, of frontal horns with the habit of rumination, are generalizations as purely empirical as those known to the aboriginal hunter. The vegetal physiologist cannot in the least understand the complex relation between the kind of leaf and the kind of fruit borne by a particular plant: he knows these and like connexions simply in the same manner that the barbarian knows them. But the fact that sundry of the uniform relations which chiefly make up the organic sciences, were very early recognized, is due to the high degree of vividness and frequency with which they were presented to consciousness. Though the connexion between the form of a given creature and the sound it makes, or the quality of its fur, or the nature of its flesh, is extremely involved; yet the two terms of the relation are conspicuous; are usually observed in close juxtaposition in time and space; are so observed perhaps daily, or many times a day; and above all a knowledge of their connexion has a direct and obvious bearing on personal welfare. Meanwhile, we see that innumerable other relations of exactly the same order, which are displayed with even greater frequency by surrounding plants and animals, remain for thousands of years unrecognized, if they are unobtrusive or of no apparent moment.

When, passing from this primitive stage to a more advanced stage, we trace the discovery of those less familiar uniformities which constitute what is technically distinguished as Science, we find the order of discovery to be still determined in the same manner. We shall most clearly see this in contemplating separately the influence of each derivative condition; as was proposed in the last section.

§ 38. How relations that have an immediate bearing on the maintenance of life, are, other things equal, necessarily fixed in the mind before those which have no such immediate bearing, is abundantly illustrated in the history of Science. The habits of existing uncivilized races, who fix times by moons and barter so many of one article for so many of another, show us that numeration, which is the germ of mathematical science, commenced under the immediate pressure of personal wants; and it can scarcely be doubted that those laws of numerical relations which are embodied in the rules of arithmetic, were first brought to light through the practice of mercantile exchange. Similarly with Geometry. The derivation shows us that it originally included only certain methods of partitioning ground and laying out buildings. The properties of the scales and the lever, involving the first principle in mechanics, were early generalized under the stimulus of commercial and architectural needs. To fix the times of religious festivals and agricultural operations, were the motives which led to the establishment of the simpler astronomic periods. Such small knowledge of chemical relations as was involved in ancient metallurgy, was manifestly obtained in seeking how to improve tools and weapons. In the alchemy of later times, we see how greatly an intense hope of private benefit contributed to the disclosure of a certain class of uniformities. Nor is our own age barren of illustrations. "Here," says Humboldt when in Guiana, "as in many parts of Europe, the sciences are thought worthy to occupy the mind, only so far as they confer some immediate and practical benefit on society." "How is it possible to believe," said a missionary to him, "that you have left your country to come and be devoured by mosquitoes on this river, and to measure lands that are not your own." Our coasts furnish like instances. Every sea-side naturalist knows how great is the contempt with which fishermen regard the collection of objects for the microscope or aquarium: their incredulity as to the possible value of such things, being so great, that they can scarcely be induced even by bribes to preserve the refuse of their nets. Nay, we need not go for evidence beyond daily table-talk. The demand for "practical science" — for a knowledge that can be brought to bear on the business of life; joined to the ridicule commonly vented on pursuits that have no obvious use; suffice to show that the order in which different coexistences and sequences are discovered, greatly depends on the directness with which they affect our welfare.

That, when all other conditions are the same, obtrusive relations will be generalized before unobtrusive ones, is so nearly a truism that examples appear almost superfluous. If it be admitted that by the aboriginal man, as by the child, the co-existent properties of large surrounding objects are noticed before those of minute objects; and that the external relations which bodies present are generalized before their internal ones; it must be admitted that in all subsequent stages of progress, the comparative conspicuousness of relations has greatly affected the order in which they were recognized as uniform. Hence it happened that after the establishment of those very manifest sequences constituting a lunation, and those less manifest ones marking a year, and those still less manifest ones marking the planetary periods, Astronomy occupied itself with such inconspicuous sequences as those displayed in the repeating cycle of lunar eclipses, and those which suggested the theory of epicycles and eccentrics; while modern Astronomy deals with still more inconspicuous sequences: some of which, as the planetary rotations, are nevertheless the simplest which the heavens present. In Physics, the early use of canoes implied an empirical knowledge of certain hydrostatic relations that are intrinsically more complex than sundry static relations then unknown; but these hydrostatic relations were thrust upon observation. Or if we compare the solution of the problem of specific gravity by Archimedes, with the discovery of atmospheric pressure by Torricelli, (the two involving mechanical relations of exactly the same kind,) we perceive that the much earlier occurrence of the first than the last, was determined neither by a difference in their bearings on personal welfare, nor by a difference in the frequency with which illustrations of them come under observation, nor by relative simplicity; but solely by the greater obtrusiveness of the connexion between antecedent and consequent in the one case than in the other. Similarly with Chemistry. The burning of wood, the rusting of iron, the putrefaction of dead bodies, were early known as consequents uniformly related to certain antecedents; but not until long after was there reached a like empirical knowledge of the effect produced by air in the decomposition of soil: a phenomenon of equal simplicity, equal or greater importance, and greater frequency; but one that is extremely unobtrusive. Among miscellaneous illustrations, it may be pointed out that the connexions between lightning and thunder and between rain and clouds, were established long before others of the same order; simply because they thrust themselves on the attention. Or the long-delayed discovery of the microscopic forms of life, with all the phenomena they present, may be named as very clearly showing how certain groups of relations that are not ordinarily perceptible, though in all other respects like long-familiar relations, have to wait until changed conditions render them perceptible. But, without further details, it needs only to consider the inquiries which now occupy the electrician, the chemist, the physiologist, to see that Science has advanced and is advancing from the more conspicuous phenomena to the less conspicuous ones.

How the degree of absolute frequency of a relation affects the recognition of its uniformity, we see in contrasting certain biological facts. Death and disease are near akin in most of their relations to us; while in respect of complexity, conspicuousness, and the directness with which they personally concern us, diseases in general may be put pretty nearly on a level with each other. But there are great differences in the times at which the natural sequences they severally exhibit are recognized as such. The connexion between death and bodily injury, constantly displayed not only in men but in all inferior creatures, was known as an established uniformity while yet diseases were thought supernatural. Among diseases themselves, it is observable that comparatively unusual ones were regarded as of demoniacal origin during ages when the more frequent were ascribed to ordinary causes: a truth paralleled indeed among our own peasantry, who by the use of charms show a lingering superstition with respect to rare disorders, which they do not show with respect to common ones, such as colds. Passing to physical illustrations, we may note that within the historic period, whirlpools were accounted for by the agency of water-spirits; but we do not find that within the same period the disappearance of water on exposure either to the sun or to artificial heat was interpreted in an analogous way: though a much more marvellous occurrence, and a much more complex one, its great frequency led to the early establishment of it as a natural uniformity. Rainbows and comets do not differ greatly in conspicuousness, and a rainbow is intrinsically the more involved phenomenon; but chiefly because of their far greater commonness, rainbows were perceived to have a direct dependence on sun and rain while yet comets were regarded as supernatural appearances.

That races living inland must long have remained ignorant of the daily and monthly sequences of the tides, and that intertropical races could not early have comprehended the phenomena of northern winters, are extreme illustrations of the influence which relative frequency has on the recognition of uniformities. Animals which, where they are indigenous, call forth no surprise by their structure or habits, because these are so familiar, when taken to a part of the earth where they have never been seen, are looked at with an astonishment approaching to awe—are even thought supernatural: a fact which will suggest numerous others that show how the localization of phenomena, in part controls the order in which they are reduced to law. Not only however does their localization in space affect the progression, but also their localization in time. Facts which are rarely if ever manifested during one era, are rendered very frequent in another, simply through the changes wrought by civilization. The lever, of which the properties are illustrated in the use of sticks and weapons, is vaguely understood by every savage— on applying it in a certain way he rightly anticipates certain effects; but the action of the equally simple wedge, which is not commonly displayed till tool-making has made some progress, is less early generalized; while the wheel and axle, pulley, and screw, cannot have their powers either empirically or rationally known till the advance of the arts has more or less familiarized them. Through those various means of exploration which we have inherited and are ever increasing, we have become acquainted with a vast range of chemical relations that were relatively non-existent to the primitive man: to highly developed industries we owe both the substances and the apparatus that have disclosed to us countless uniformities which our ancestors had no opportunity of seeing, and therefore could not recognize. These and sundry like instances that will occur to the reader, show that the accumulated materials, and processes, and appliances, and products, which characterize the environments of complex societies, greatly increase the accessibility of various classes of relations; and by so multiplying the experiences of them, or making them relatively frequent, facilitate their generalization. To which add, that various classes of phenomena presented by society itself, as for instance those which political economy formulates, become relatively frequent and therefore recognizable in advanced social states; while in less advanced ones they are too rarely displayed to have their relations perceived, or, as in the least advanced ones, are not displayed at all.

That, where no other circumstances interfere, the order in which different uniformities are established varies as their complexity, is manifest. The geometry of straight lines was understood before the geometry of curved lines; the properties of the circle before the properties of the ellipse, parabola and hyperbola; and the equations of curves of single curvature were ascertained before those of curves of double curvature. Plane trigonometry comes in order of time and simplicity before spherical trigonometry; and the mensuration of plane surfaces and solids before the mensuration of curved surfaces and solids. Similarly with mechanics: the laws of simple motion were generalized before those of compound motion; and those of rectilinear motion before those of curvilinear motion. The properties of equal-armed levers, or scales, were understood before those of the lever with unequal arms; and the law of the inclined plane was formulated earlier than that of the screw, which involves it. In chemistry, the progress has been from the simple inorganic compounds, to the more involved organic ones. And where, as in most of the other sciences, the conditions of the exploration are more complicated, we still may clearly trace relative complexity as one of the determining circumstances.

The progression from concrete relations to abstract ones, and from the less abstract to the more abstract, is equally obvious. Numeration, which in its primary form concerned itself only with groups of actual objects, came earlier than simple arithmetic: the rules of which deal with numbers apart from objects. Arithmetic, limited in its sphere to concrete numerical relations, is alike earlier and less abstract than Algebra, which deals with the relations of these relations. And in like manner, the Infinitesimal Calculus comes after Algebra, both in order of evolution and in order of abstractness. In Astronomy, the progress has been from special generalizations, each expressing the motions of a particular planet, to the generalizations of Kepler, expressing the motions of the planets at large; and then to Newton's generalization, expressing the motions of all heavenly bodies whatever. Similarly with Physics, Chemistry and Biology, there has ever been an advance from the relations of particular facts and particular classes of facts, to the relations presented by still wider classes—to truths of a high generality or greater abstractness.

Brief and rude as is this sketch of a mental development that has been long and complicated, it fulfils its end if it displays the several conditions that have regulated the course of the development. I venture to think it shows inductively, what was deductively inferred, that the order in which separate groups of uniformities are recognized, depends not on one circumstance but on several circumstances. A survey of the facts makes it manifest that the various classes of relations are generalized in a certain succession, not solely because of one particular kind of difference in their natures; but also because they are variously placed with respect to time, space, other relations, and our own constitutions: our perception of them being influenced by all these conditions in endless combinations. The comparative degrees of importance, of obtrusiveness, of absolute frequency, of relative frequency, of simplicity, of concreteness, are every one of them factors; and from their union in proportions that are more or less different in every case, there results a highly complex process of mental evolution. But while it thus becomes manifest that the proximate causes of the succession in which relations are reduced to law, are numerous and involved; it also becomes manifest that there is one ultimate cause to which these proximate ones are subordinate. As the several circumstances that determine the early or late recognition of uniformities, are circumstances that determine the number and strength of the impressions which these uniformities make on the mind; it follows that the progression conforms to a certain fundamental principle of psychology. We see *à posteriori*, what we concluded *à priori*, that the order in which relations are generalized, depends on the frequency and impressiveness with which they are repeated in conscious experience.

§ 39. And now to observe the bearings of these truths on our general argument. Having roughly analyzed the progress of the past, let us take advantage of the light thus thrown on the present, and consider what is implied respecting the future.

Note first that the likelihood of the universality of Law, has been ever growing greater. Out of the countless coexistences and sequences with which mankind are environed, they have been continually transferring some from the group whose order was supposed to be arbitrary, to the group whose order is known to be uniform. Age by age, the number of recognized connexions of phenomena has been increasing; and that of unrecognized connexions decreasing. And manifestly, as fast as the class of ungeneralized relations becomes smaller, the probability that among them there may be some that do not conform to law, becomes less. To put the argument numerically—It is clear that when out of surrounding phenomena a hundred of several kinds have been found to occur in constant connexions, there arises a slight presumption that all phenomena occur in constant connexions. When uniformity has been established in a thousand cases, more varied in their kinds, the presumption gains strength. And when the established cases of uniformity mount to myriads, including many of each variety, it becomes an ordinary induction that uniformity exists everywhere. Just as from the numerous observed cases in which heavenly bodies have been found to move in harmony with the law of gravitation, it is inferred that all heavenly bodies move in harmony with the law of gravitation; so, from the innumerable observed cases in which phenomena are found to stand in invariable connexions, it is inferred that in all cases phenomena stand in invariable connexions.

Silently and insensibly their experiences have been pressing men on towards the conclusion thus drawn. Not out of a conscious regard for these abstract reasons, but from a habit of thought which these abstract reasons formulate and justify, all minds have been advancing towards a belief in the constancy of surrounding coexistences and sequences. Familiarity with special uniformities, has generated the abstract conception of uniformity—the idea of *Law*; and this idea has been in successive generations slowly gaining fixity and clearness. Especially has it been thus among those whose knowledge of natural phenomena is the most extensive—men of science. The Mathematician, the Physicist, the Astronomer, the Chemist, severally acquainted with the vast accumulations of uniformities established by their predecessors, and themselves daily adding new ones as well as verifying the old, acquire a far stronger faith in Law than is ordinarily possessed. With them this faith, ceasing to be merely passive, becomes an active stimulus to inquiry. Wherever there exist phenomena of which the dependence is not yet ascertained, these most cultivated intellects, impelled by the conviction that here too there is some invariable connexion, proceed to observe, compare, and experiment; and when they discover the law to which the phenomena conform, as they eventually do, their general belief in the universality of law is further strengthened. So overwhelming is the evidence, and such the effect of this discipline, that to the advanced student of nature, the proposition that there are lawless phenomena, has become not only incredible but almost inconceivable.

Hence we may see how inevitably there must spread among mankind at large, this habitual recognition of law which already distinguishes modern thought from ancient thought. Not only is it that each conquest of generalization over a region of fact hitherto ungeneralized, and each merging of lower generalizations in a higher one, adds to the distinctness of this recognition among those immediately concerned—not only is it that the fulfilment of the predictions made possible by every new step, and the further command so gained of nature's forces, prove to the uninitiated the validity of these generalizations and the doctrine they illustrate; but it is that widening education is daily diffusing among the mass of men, that knowledge of generalizations which has been hitherto confined to the few. And as fast as this diffusion goes on, must the belief of the scientific become the belief of the world at large. The simple accumulation of instances, must inevitably establish in the general mind, a conviction of the universality of law; even were the influence of this accumulation to be aided by no other.

§ 40. But it will be aided by another. From the evidence above set forth, it may be inferred that a secondary influence will by and by enforce this primary one. That law is universal, will become an irresistible conclusion when it is perceived that *the progress in the discovery of laws itself conforms to law;* and when it is hence understood why certain groups of phenomena have been reduced to law, while other groups are still unreduced. When it is seen that the order in which uniformities are recognized, must depend upon the frequency and vividness with which they are repeated in conscious experience; when it is seen that, as a matter of fact, the most common, important, conspicuous, concrete and simple uniformities were the earliest recognized, because they were experienced oftenest and most distinctly; when it is further seen that from the beginning the advance has been to the recognition of uniformities which, from one or other circumstance, were less often experienced; it will by implication be seen that long after the great mass of phenomena have been generalized, there must remain phenomena which, from their rareness, or unobtrusiveness, or seeming unimportance, or complexity, or abstractness, are still ungeneralized. Thus will be furnished a solution to a difficulty sometimes raised. When it is asked why the universality of law is not already fully established, there will be the answer that the directions in which it is not yet established are those in which its establishment must necessarily be latest. That state of things which is inferable beforehand, is just the state which we find to exist. If such coexistences and sequences as those of Biology and Sociology are not yet reduced to law, the presumption is not that they are irreducible to law, but that their laws elude our present means of analysis. Having long ago proved uniformity throughout all the lower classes of relations; and having been step by step proving uniformity throughout classes of relations successively higher and higher; if we have not at present succeeded with the highest classes, it may be fairly concluded that our powers are at fault, rather than that the uniformity does not exist. And unless we make the absurd assumption that the process of generalization, now going on with unexampled rapidity, has reached its limit, and will suddenly cease, we must infer that ultimately mankind will discover a constant order of manifestation even in the most involved, obscure, and abstract phenomena.

§ 41. Not even yet, however, have we exhausted the evidence. The foregoing arguments have to be merged in another, still more cogent, which fuses all fragmentary proofs into one general proof.

Thus far we have spoken of laws that are more or less special; and from the still-continuing disclosure of special laws, each formulating some new class of phenomena, have inferred that eventually all classes of phenomena will be formulated. If, now, we find that there are laws of far higher generality, to which those constituting the body of Science are subordinate; the fact must greatly strengthen the proof that Law is universal. If, underneath different groups of concrete phenomena, Mechanical, Chemical, Thermal, Electric, &c., we discern certain uniformities of action common to them all; we have a new and weighty reason for believing that uniformity of action pervades the whole of nature. And if we also see that these most general laws hold not only of the inorganic but of the organic worlds—if we see that the phenomena of Life, of Mind, of Society, whose special laws are yet unestablished, nevertheless conform to these most general laws; the proof of the universality of Law amounts to demonstration.

That there are laws of this transcendant generality, has now to be shown. To specify and illustrate them, will be the purpose of the succeeding chapters. And while, in contemplating them, we shall perceive how irresistible is the conclusion that the workings of the Unknowable are distinguished from those of finite agents by their absolute uniformity; we shall at the same time familiarize ourselves with those primary facts through which all other facts are to be interpreted.

CHAPTER II
THE LAW OF EVOLUTION[8]

§ 42. The class of phenomena to be considered under the title of Evolution, is in a great measure co-extensive with the class commonly indicated by the word Progress. But the word Progress is here inappropriate, for several reasons. To specify these reasons will perhaps be the best way of showing what is to be understood by Evolution.

In the first place, the current conception of Progress is shifting and indefinite. Sometimes it comprehends little more than simple growth—as of a nation in the number of its members and the extent of territory over which it has spread. At other times it has reference to quantity of material products—as when the advance of agriculture and manufactures is the topic. Now the superior quality of these products is contemplated; and then the new or improved appliances by which they are produced. When, again, we speak of moral or intellectual progress, we refer to the state of the individual or people exhibiting it; while, when the progress of Knowledge, of Science, of Art, is commented upon, we have in view certain abstract results of human thought and action. In the second place, besides being more or less vague, the ordinary idea of Progress is in great measure erroneous. It takes in not so much the reality as its accompaniments—not so much the substance as the shadow. That progress in intelligence seen during the growth of the child into the man, or the savage into the philosopher, is commonly regarded as consisting in the greater number of facts known and laws understood; whereas the actual progress consists in those internal modifications of which this increased knowledge is the expression. Social progress is supposed to consist in the produce of a greater quantity and variety of the articles required for satisfying men's wants—in the increasing security of person and property—in widening freedom of action; whereas, rightly understood, social progress consists in those changes of structure in the social organism which have entailed these consequences. The interpretation is a teleological one. The phenomena are contemplated solely as bearing on human happiness. Only those changes are held to constitute progress, which directly or indirectly tend to heighten human happiness. And they are thought to constitute progress simply *because* they tend to heighten human happiness. In the third place, in consequence of its teleological implications, the term Progress is rendered scarcely applicable to a wide range of phenomena which are intrinsically of the same nature as those included under it. The metamorphoses of an insect are only by analogy admitted within the scope of the word, as popularly accepted; though, considered in themselves, they have as much right there as the changes which constitute civilization. Having no apparent bearing on human interests, an increasing complication in the arrangement of ocean-currents, would not ordinarily be regarded as progress; though really of the same character as phenomena which are so regarded.

Hence the need for another word. Our purpose here is to analyze the various class of changes usually considered as Progress, together with others like them which are not so considered; and to see what is their intrinsic peculiarity—what is their essential nature apart from their bearings on our welfare. And that we may avoid the confusion of thought likely to result from pre-established associations, it will be best to substitute for the term Progress, the term Evolution. Our question is then—what is Evolution?

§ 43. In respect to that evolution which individual organisms display, this question has been answered. Pursuing an idea which Harvey set afloat, Wolff, Goethe, and Von Baer, have established the truth that the series of changes gone through during the development of a seed into a tree, or an ovum into an animal, constitute an advance from homogeneity of structure to heterogeneity of structure. In its primary stage, every germ consists of a substance that is uniform throughout, both in texture and chemical composition. The first step is the appearance of a difference between two parts of this substance; or, as the phenomenon is called in physiological language, a differentiation. Each of these differentiated divisions presently begins itself to exhibit some contrast of parts; and by and by these secondary differentiations become as definite as the original one. This process is continuously repeated—is simultaneously going on in all parts of the growing embryo; and by endless such differentiations there is finally produced that complex combination of tissues and organs, constituting the adult animal or plant. This is the history of all organisms whatever. It is settled beyond dispute that organic evolution consists in a change from the homogeneous to the heterogeneous.

Now I propose in the first place to show, that this law of organic evolution is the law of all evolution. Whether it be in the development of the Earth, in the development of Life upon its surface, in the development of Society, of Government, of Manufactures, of Commerce, of Language, Literature, Science, Art, this same advance from the simple to the complex, through successive differentiations, holds uniformly. From the earliest traceable cosmical changes down to the latest results of civilization, we shall find that the transformation of the homogeneous into the heterogeneous, is that in which Evolution essentially consists.

§ 44. With the view of showing that *if* the Nebular Hypothesis be true, the genesis of the solar system supplies one illustration of this law, let us assume that the matter of which the sun and planets consist was once in a diffused form; and that from the gravitation of its atoms there resulted a gradual concentration. By the hypothesis, the solar system in its nascent state existed as an indefinitely extended and nearly homogeneous medium—a medium almost homogeneous in density, in temperature, and in other physical attributes. The first advance towards consolidation resulted in a differentiation between the occupied space which the nebulous mass still filled, and the unoccupied space which it previously filled. There simultaneously resulted a contrast in density and a contrast in temperature, between the interior and the exterior of this mass. And at the same time there arose throughout it, rotatory movements, whose velocities varied according to their distances from its centre. These differentiations increased in number and degree until there was evolved the organized group of sun, planets, and satellites, which we now know—a group which presents numerous contrasts of structure and action among its members. There are the immense contrasts between the sun and the planets, in bulk and in weight; as well as the subordinate contrasts between one planet and another, and between the planets and their satellites. There is the similarly marked contrast between the sun as almost stationary, and the planets as moving round him with great velocity; while there are the secondary contrasts between the velocities and periods of the several planets, and between their simple revolutions and the double ones of their satellites, which have to move round their primaries while moving round the sun. There is the yet further strong contrast between the sun and the planets in respect of temperature; and there is reason to suppose that the planets and satellites differ from each other in their proper heat, as well as in the heat they receive from the sun. When we bear in mind that, in addition to these various contrasts, the planets and satellites also differ in respect to their distances from each other and their primary; in respect to the inclinations of their orbits, the inclinations of their axes, their times of rotation on their axes, their specific gravities, and their physical constitutions; we see what a high degree of heterogeneity the solar system exhibits, when compared with the almost complete homogeneity of the nebulous mass out of which it is supposed to have originated.

§ 45. Passing from this hypothetical illustration, which must be taken for what it is worth, without prejudice to the general argument, let us descend to a more certain order of evidence.

It is now generally agreed among geologists that the Earth was at first a mass of molten matter; and that it is still fluid and incandescent at the distance of a few miles beneath its surface. Originally, then, it was homogeneous in consistence, and, because of the circulation that takes place in heated fluids, must have been comparatively homogeneous in temperature; and it must have been surrounded by an atmosphere consisting partly of the elements of air and water, and partly of those various other elements which assume a gaseous form at high temperatures. That slow cooling by radiation which is still going on at an inappreciable rate, and which, though originally far more rapid than now, necessarily required an immense time to produce any decided change, must ultimately have resulted in the solidification of the portion most able to part with its heat; namely, the surface. In the thin crust thus formed, we have the first marked differentiation. A still further cooling, a consequent thickening of this crust, and an accompanying deposition of all solidifiable elements contained in the atmosphere, must finally have been followed by the condensation of the water previously existing as vapour. A second marked differentiation must thus have arisen; and as the condensation must have taken place on the coolest parts of the surface—namely, about the poles—there must thus have resulted the first geographical distinction of parts.

To these illustrations of growing heterogeneity, which, though deduced from the known laws of matter, may be regarded as more or less hypothetical, Geology adds an extensive series that have been inductively established. Its investigations show that the Earth has been continually becoming more heterogeneous through the multiplication of the strata which form its crust; further, that it has been becoming more heterogeneous in respect of the composition of these strata, the latter of which, being made from the detritus of the older ones, are many of them rendered highly complex by the mixture of materials they contain; and that this heterogeneity has been vastly increased by the action of the Earth's still molten nucleus upon its envelope: whence have resulted not only a great variety of igneous rocks, but the tilting up of sedimentary strata at all angles, the formation of faults and metallic veins, the production of endless dislocations and irregularities. Yet again, geologists teach us that the Earth's surface has been growing more varied in elevation—that the most ancient mountain systems are the smallest, and the Andes and Himalayas the most modern; while, in all probability, there have been corresponding changes in the bed of the ocean. As a consequence of these ceaseless differentiations, we now find that no considerable portion of the Earth's exposed surface is like any other portion, either in contour, in geologic structure, or in chemical composition; and that in most parts it changes from mile to mile in all these characteristics.

Moreover, it must not be forgotten that there has been simultaneously going on a gradual differentiation of climates. As fast as the Earth cooled and its crust solidified, there arose appreciable differences in temperature between those parts of its surface most exposed to the sun and those less exposed. Gradually, as the cooling progressed, these differences became more pronounced; until there finally resulted the marked contrasts between regions of perpetual ice and snow, regions where winter and summer alternately reign for periods varying according to the latitude, and regions where summer follows summer with scarcely an appreciable variation. At the same time, the successive elevations and subsidences of different portions of the Earth's crust, tending as they have done to the present irregular distribution of land and sea, have entailed various modifications of climate beyond those dependent on latitude; while a yet further series of such modifications have been produced by increasing differences of elevation in the land, which have in sundry places brought arctic, temperate, and tropical climates to within a few miles of each other. And the general result of these changes is, that not only has every extensive region its own meteorologic conditions, but that every locality in each region differs more or less from others in those conditions: as in its structure, its contour, its soil.

Thus, between our existing Earth, the phenomena of whose varied crust neither geographers, geologists, mineralogists nor meteorologists have yet enumerated, and the molten globe out of which it was evolved, the contrast in heterogeneity is sufficiently striking.

§ 46. When from the Earth itself we turn to the plants and animals that have lived, or still live, upon its surface, we find ourselves in some difficulty from lack of facts. That every existing organism has been developed out of the simple into the complex, is indeed the first established truth of all; and that every organism which has existed was similarly developed, is an inference that no physiologist will hesitate to draw. But when we pass from individual forms of life to Life in general, and inquire whether the same law is seen in the *ensemble* of its manifestations,—whether modern plants and animals are of more heterogeneous structure than ancient ones, and whether the Earth's present Flora and Fauna are more heterogeneous than the Flora and Fauna of the past,—we find the evidence so fragmentary, that every conclusion is open to dispute. Two-thirds of the Earth's surface being covered by water; a great part of the exposed land being inaccessible to, or untravelled by, the geologist; the greater part of the remainder having been scarcely more than glanced at; and even the most familiar portions, as England, having been so imperfectly explored, that a new series of strata has been added within these few years,—it is manifestly impossible for us to say with any certainty what creatures have, and what have not, existed at any particular period. Considering the perishable nature of many of the lower organic forms, the metamorphosis of many sedimentary strata, and the gaps that occur among the rest, we shall see further reason for distrusting our deductions. On the one hand, the repeated discovery of vertebrate remains in strata previously supposed to contain none,—of reptiles where only fish were thought to exist,—of mammals where it was believed there were no creatures higher than reptiles; renders it daily more manifest how small is the value of negative evidence. On the other hand, the worthlessness of the assumption that we have discovered the earliest, or anything like the earliest, organic remains, is becoming equally clear. That the oldest known aqueous formations have been greatly changed by igneous action, and that still older ones have been totally transformed by it, is becoming undeniable. And the fact that sedimentary strata earlier than any we know, have been melted up, being admitted, it must also be admitted that we cannot say how far back in time this destruction of sedimentary strata has been going on. Thus it is manifest that the title *Palæozoic*, as applied to the earliest known fossiliferous strata, involves a *petitio principii*; and that, for aught we know to the contrary, only the last few chapters of the Earth's biological history may have come down to us.

All inferences drawn from such scattered facts as we find, must thus be extremely questionable. If, looking at the general aspect of evidence, a progressionist argues that the earliest known vertebrate remains are those of Fishes, which are the most homogeneous of the vertebrata; that Reptiles, which are more heterogeneous, are later; and that later still, and more heterogeneous still, are Mammals and Birds; it may be replied that the Palæozoic deposits, not being estuary deposits, are not likely to contain the remains of terrestrial vertebrata, which may nevertheless have existed at that era. The same answer may be made to the argument that the vertebrate fauna of the Palæozoic period, consisting so far as we know, entirely of Fishes, was less heterogeneous than the modern vertebrate fauna, which includes Reptiles, Birds and Mammals, of multitudinous genera; or the uniformitarian may contend with great show of truth, that this appearance of higher and more varied forms in later geologic eras, was due to progressive immigration—that a continent slowly upheaved from the ocean at a point remote from pre-existing continents, would necessarily be peopled from them in a succession like that which our strata display. At the same time the counter-arguments may be proved equally inconclusive. When, to show that there cannot have been a continuous evolution of the more homogeneous organic forms into the more heterogeneous ones, the uniformitarian points to the breaks that occur in the succession of these forms; there is the sufficient answer that current geological changes show us why such breaks must occur, and why, by subsidences and elevations of large area, there must be produced such marked breaks as those which divide the three great geologic epochs. Or again, if the opponent of the development hypothesis cites the facts set forth by Professor Huxley in his lecture on "Persistent Types"—if he points out that "of some two hundred known orders of plants, not one is exclusively fossil," while "among animals, there is not a single totally extinct class; and of the orders, at the outside not more than seven per cent. are unrepresented in the existing creation"—if he urges that among these some have continued from the Silurian epoch to our own day with scarcely any change—and if he infers that there is evidently a much greater average resemblance between the living forms of the past and those of the present, than consists with this hypothesis; there is still a satisfactory reply, on which in fact Prof. Huxley insists; namely, that we have evidence of a "pre-geologic era" of unknown duration. And indeed, when it is remembered, that the enormous subsidences of the Silurian period show the Earth's crust to have been approximately as thick then as it is now—when it is concluded that the time taken to form so thick a crust, must have been immense as compared with the time which has since elapsed—when it is assumed, as it must be, that during this comparatively immense time the geologic and biologic changes went on at their usual rates; it becomes manifest, not only that the palæontological records which we find, do not negative the theory of evolution, but that they are such as might rationally be looked for.

Moreover, it must not be forgotten that though the evidence suffices neither for proof nor disproof, yet some of its most conspicuous facts support the belief, that the more heterogeneous organisms and groups of organisms, have been evolved from the less heterogeneous ones. The average community of type between the fossils of adjacent strata, and still more the community that is found between the latest tertiary fossils and creatures now existing, is one of these facts. The discovery in some modern deposits of such forms as the Palæotherium and Anaplotherium, which, if we may rely on Prof. Owen, had a type of structure intermediate between some of the types now existing, is another of these facts. And the comparatively recent appearance of Man, is a third fact of this kind, which possesses still greater significance. Hence we may say, that though our knowledge of past life upon the Earth, is too scanty to justify us in asserting an evolution of the simple into the complex, either in individual forms or in the aggregate of forms; yet the knowledge we have, not only consists with the belief that there has been such an evolution, but rather supports it than otherwise.

§ 47. Whether an advance from the homogeneous to the heterogeneous is or is not displayed in the biological history of the globe, it is clearly enough displayed in the progress of the latest and most heterogeneous creature—Man. It is alike true that, during the period in which the Earth has been peopled, the human organism has grown more heterogeneous among the civilized divisions of the species; and that the species, as a whole, has been made more heterogeneous by the multiplication of races and the differentiation of these races from each other. In proof of the first of these positions, we may cite the fact that, in the relative development of the limbs, the civilized man departs more widely from the general type of the placental mammalia, than do the lower human races. Though often possessing well-developed body and arms, the Papuan has extremely small legs: thus reminding us of the quadrumana, in which there is no great contrast in size between the hind and fore limbs. But in the European, the greater length and massiveness of the legs has become very marked—the fore and hind limbs are relatively more heterogeneous. Again, the greater ratio which the cranial bones bear to the facial bones, illustrates the same truth. Among the vertebrata in general, evolution is marked by an increasing heterogeneity in the vertebral column, and more especially in the segments constituting the skull: the higher forms being distinguished by the relatively larger size of the bones which cover the brain, and the relatively smaller size of those which form the jaws, &c. Now, this characteristic, which is stronger in Man than in any other creature, is stronger in the European than in the savage. Moreover, judging from the greater extent and variety of faculty he exhibits, we may infer that the civilized man has also a more complex or heterogeneous nervous system than the uncivilized man; and indeed the fact is in part visible in the increased ratio which his cerebrum bears to the subjacent ganglia. If further elucidation be needed, we may find it in every nursery. The infant European has sundry marked points of resemblance to the lower human races; as in the flatness of the alæ of the nose, the depression of its bridge, the divergence and forward opening of the nostrils, the form of the lips, the absence of a frontal sinus, the width between the eyes, the smallness of the legs. Now, as the developmental process by which these traits are turned into those of the adult European, is a continuation of that change from the homogeneous to the heterogeneous displayed during the previous evolution of the embryo, which every physiologist will admit; it follows that the parallel developmental process by which the like traits of the barbarous races have been turned into those of the civilized races, has also been a continuation of the change from the homogeneous to the heterogeneous. The truth of the second position—that Mankind, as a whole, have become more heterogeneous—is so obvious as scarcely to need illustration. Every work on Ethnology, by its divisions and subdivisions of races, bears testimony to it. Even were we to admit the hypothesis that Mankind originated from several separate stocks, it would still remain true that as, from each of these stocks, there have sprung many now widely different tribes, which are proved by philological evidence to have had a common origin, the race as a whole is far less homogeneous than it once was. Add to which, that we have, in the Anglo-Americans, an example of a new variety arising within these few generations; and that, if we may trust to the descriptions of observers, we are likely soon to have another such example in Australia.

§ 48. On passing from Humanity under its individual form, to Humanity as socially embodied, we find the general law still more variously exemplified. The change from the homogeneous to the heterogeneous, is displayed equally in the progress of civilization as a whole, and in the progress of every tribe or nation; and is still going on with increasing rapidity.

As we see in existing barbarous tribes, society in its first and lowest form is a homogeneous aggregation of individuals having like powers and like functions: the only marked difference of function being that which accompanies difference of sex. Every man is warrior, hunter, fisherman, tool-maker, builder; every woman performs the same drudgeries; every family is self-sufficing, and, save for purposes of aggression and defence, might as well live apart from the rest. Very early, however, in the process of social evolution, we find an incipient differentiation between the governing and the governed. Some kind of chieftainship seems coeval with the first advance from the state of separate wandering families to that of a nomadic tribe. The authority of the strongest makes itself felt among a body of savages, as in a herd of animals, or a posse of school-boys. At first, however, it is indefinite, uncertain; is shared by others of scarcely inferior power; and is unaccompanied by any difference in occupation or style of living: the first ruler kills his own game, makes his own weapons, builds his own hut, and, economically considered, does not differ from others of his tribe. Gradually, as the tribe progresses, the contrast between the governing and the governed grows more decided. Supreme power becomes hereditary in one family; the head of that family, ceasing to provide for his own wants, is served by others; and he begins to assume the sole office of ruling. At the same time there has been arising a co-ordinate species of government— that of Religion. As all ancient records and traditions prove, the earliest rulers are regarded as divine personages. The maxims and commands they uttered during their lives are held sacred after their deaths, and are enforced by their divinely-descended successors; who in their turns are promoted to the pantheon of the race, there to be worshipped and propitiated along with their predecessors: the most ancient of whom is the supreme god, and the rest subordinate gods. For a long time these connate forms of government—civil and religious—continue closely associated. For many generations the king continues to be the chief priest, and the priesthood to be members of the royal race. For many ages religious law continues to contain more or less of civil regulation, and civil law to possess more or less of religious sanction; and even among the most advanced nations these two controlling agencies are by no means completely differentiated from each other. Having a common root with these, and gradually diverging from them, we find yet another controlling agency—that of Manners or ceremonial usages. All titles of honour are originally the names of the god-king; afterwards of God and the king; still later of persons of high rank; and finally come, some of them, to be used between man and man. All forms of complimentary address were at first the expressions of submission from prisoners to their conqueror, or from subjects to their ruler, either human or divine—expressions that were afterwards used to propitiate subordinate authorities, and slowly descended into ordinary intercourse. All modes of salutation were once obeisances made before the monarch and used in worship of him after his death. Presently others of the god-descended race were similarly saluted; and by degrees some of the salutations have become the due of all.[9] Thus, no sooner does the originally homogeneous social mass differentiate into the governed and the governing parts, than this last exhibits an incipient differentiation into religious and secular—Church and State; while at the same time there begins to be differentiated from both, that less definite species of government which rides our daily intercourse—a species of government which, as we may see in heralds' colleges, in books of the peerage, in masters of ceremonies, is not without a certain embodiment of its own. Each of these kinds of government is itself subject to successive differentiations. In the course of ages, there arises, as among ourselves, a highly complex political organization of monarch, ministers, lords and commons, with their subordinate administrative departments, courts of justice, revenue offices, &c., supplemented in the provinces by municipal governments, county governments, parish or union governments—all of them more or less elaborated. By its side there grows up a highly complex religious organization, with its various grades of officials from archbishops down to sextons, its colleges, convocations, ecclesiastical courts, &c.; to all which must be added the ever-multiplying independent sects, each with its general and local authorities. And at the same time there is developed a highly complex aggregation of customs, manners, and temporary fashions, enforced by society at large, and serving to control those minor transactions between man and man which are not regulated by civil and religious law. Moreover, it is to be observed that this ever-increasing heterogeneity in the governmental appliances of each nation, has been accompanied by an increasing heterogeneity in the governmental appliances of different nations: all of which are more or less unlike in their political systems and legislation, in their creeds and religious institutions, in their customs and ceremonial usages.

Simultaneously there has been going on a second differentiation of a more familiar kind; that, namely, by which the mass of the community has been segregated into distinct classes and orders of workers. While the governing part has undergone the complex development above detailed, the governed part has undergone an equally complex development; which has resulted in that minute division of labour characterizing advanced nations. It is needless to trace out this progress from its first stages, up through the caste divisions of the East and the incorporated guilds of Europe, to the elaborate producing and distributing organization existing among ourselves. Political economists have long since indicated the evolution which, beginning with a tribe whose members severally perform the same actions, each for himself ends with a civilized community whose members severally perform different actions for each other; and they have further pointed out the changes through which the solitary producer of any one commodity, is transformed into a combination of producers who, united under a master, take separate parts in the manufacture of such commodity. But there are yet other and higher phases of this advance from the homogeneous to the heterogeneous in the industrial organization of society. Long after considerable progress has been made in the division of labour among the different classes of workers, there is still little or no division of labour among the widely separated parts of the community: the nation continues comparatively homogeneous in the respect that in each district the same occupations are pursued. But when roads and other means of transit become numerous and good, the different districts begin to assume different functions, and to become mutually dependent. The calico-manufacture locates itself in this county, the woollen-manufacture in that; silks are produced here, lace there; stockings in one place, shoes in another; pottery, hardware, cutlery, come to have their special towns; and ultimately every locality grows more or less distinguished from the rest by the leading occupation carried on in it. Nay, more, this subdivision of functions shows itself not only among the different parts of the same nation, but among different nations. That exchange of commodities which free-trade promises so greatly to increase, will ultimately have the effect of specializing, in a greater or less degree, the industry of each people. So that beginning with a barbarous tribe, almost if not quite homogeneous in the functions of its members, the progress has been, and still is, towards an economic aggregation of the whole human race; growing ever more heterogeneous in respect of the separate functions assumed by separate nations, the separate functions assumed by the local sections of each nation, the separate functions assumed by the many kinds of makers and traders in each town, and the separate functions assumed by the workers united in producing each commodity.

§ 49. Not only is the law thus clearly exemplified in the evolution of the social organism, but it is exemplified with equal clearness in the evolution of all products of human thought and action; whether concrete or abstract, real or ideal. Let us take Language as our first illustration.

The lowest form of language is the exclamation, by which an entire idea is vaguely conveyed through a single sound; as among the lower animals. That human language ever consisted solely of exclamations, and so was strictly homogeneous in respect of its parts of speech, we have no evidence. But that language can be traced down to a form in which nouns and verbs are its only elements, is an established fact. In the gradual multiplication of parts of speech out of these primary ones—in the differentiation of verbs into active and passive, of nouns into abstract and concrete—in the rise of distinctions of mood, tense, person, of number and case—in the formation of auxiliary verbs, of adjectives, adverbs, pronouns, prepositions, articles— in the divergence of those orders, genera, species, and varieties of parts of speech by which civilized races express minute modifications of meaning— we see a change from the homogeneous to the heterogeneous. And it may be remarked, in passing, that it is more especially in virtue of having carried this subdivision of functions to a greater extent and completeness, that the English language is superior to all others. Another aspect under which we may trace the development of language, is the differentiation of words of allied meanings. Philology early disclosed the truth that in all languages words may be grouped into families having a common ancestry. An aboriginal name, applied indiscriminately to each of an extensive and ill-defined class of things or actions, presently undergoes modifications by which the chief divisions of the class are expressed. These several names springing from the primitive root, themselves become the parents of other names still further modified. And by the aid of those systematic modes which presently arise, of making derivatives and forming compound terms expressing still smaller distinctions, there is finally developed a tribe of words so heterogeneous in sound and meaning, that to the uninitiated it seems incredible they should have had a common origin. Meanwhile, from other roots there are being evolved other such tribes, until there results a language of some sixty thousand or more unlike words, signifying as many unlike objects, qualities, acts. Yet another way in which language in general advances from the homogeneous to the heterogeneous, is in the multiplication of languages. Whether, as Max Müller and Bunsen think, all languages have grown from one stock, or whether, as some philologists say, they have grown from two or more stocks, it is clear that since large families of languages, as the Indo-European, are of one parentage, they have become distinct through a process of continuous divergence. The same diffusion over the Earth's surface which has led to the differentiation of the race, has simultaneously led to a differentiation of their speech: a truth which we see further illustrated in each nation by the peculiarities of dialect found in separate districts. Thus the progress of Language conforms to the general law, alike in the evolution of languages, in the evolution of families of words, and in the evolution of parts of speech.

On passing from spoken to written language, we come upon several classes of facts, all having similar implications. Written language is connate with Painting and Sculpture; and at first all three are appendages of Architecture, and have a direct connexion with the primary form of all Government—the theocratic. Merely noting by the way the fact that sundry wild races, as for example the Australians and the tribes of South Africa, are given to depicting personages and events upon the walls of caves, which are probably regarded as sacred places, let us pass to the case of the Egyptians. Among them, as also among the Assyrians, we find mural paintings used to decorate the temple of the god and the palace of the king (which were, indeed, originally identical); and as such they were governmental appliances in the same sense that state-pageants and religious feasts were. Further, they were governmental appliances in virtue of representing the worship of the god, the triumphs of the god-king, the submission of his subjects, and the punishment of the rebellious. And yet again they were governmental, as being the products of an art reverenced by the people as a sacred mystery. From the habitual use of this pictorial representation, there naturally grew up the but slightly-modified practice of picture-writing—a practice which was found still extant among the Mexicans at the time they were discovered. By abbreviations analogous to those still going on in our own written and spoken language, the most familiar of these pictured figures were successively simplified; and ultimately there grew up a system of symbols, most of which had but a distant resemblance to the things for which they stood. The inference that the hieroglyphics of the Egyptians were thus produced, is confirmed by the fact that the picture-writing of the Mexicans was found to have given birth to a like family of ideographic forms; and among them, as among the Egyptians, these had been partially differentiated into the *kuriological* or imitative, and the *tropical* or symbolic: which were, however, used together in the same record. In Egypt, written language underwent a further differentiation; whence resulted the *hieratic* and the *epistolographic* or *enchorial*: both of which are derived from the original hieroglyphic. At the same time we find that for the expression of proper names, which could not be otherwise conveyed, phonetic symbols were employed; and though it is alleged that the Egyptians never actually achieved complete alphabetic writing, yet it can scarcely be doubted that these phonetic symbols occasionally used in aid of their ideographic ones, were the germs out of which alphabetic writing grew. Once having become separate from hieroglyphics, alphabetic writing itself underwent numerous differentiations—multiplied alphabets were produced: between most of which, however, more or less connexion can still be traced. And in each civilized nation there has now grown up, for the representation of one set of sounds, several sets of written signs, used for distinct purposes. Finally, through a yet more important differentiation came printing; which, uniform in kind as it was at first, has since become multiform.

§ 50. While written language was passing through its earlier stages of development, the mural decoration which formed its root was being differentiated into Painting and Sculpture. The gods, kings, men, and animals represented, were originally marked by indented outlines and coloured. In most cases these outlines were of such depth, and the object they circumscribed so far rounded and marked out in its leading parts, as to form a species of work intermediate between intaglio and bas-relief. In other cases we see an advance upon this: the raised spaces between the figures being chiselled off, and the figures themselves appropriately tinted, a painted bas-relief was produced. The restored Assyrian architecture at Sydenham, exhibits this style of art carried to greater perfection—the persons and things represented, though still barbarously coloured, are carved out with more truth and in greater detail; and in the winged lions and bulls used for the angles of gateways, we may see a considerable advance towards a completely sculptured figure; which, nevertheless, is still coloured, and still forms part of the building. But while in Assyria the production of a statue proper, seems to have been little, if at all, attempted, we may trace in Egyptian art the gradual separation of the sculptured figure from the wall. A walk through the collection in the British Museum will clearly show this; while it will at the same time afford an opportunity of observing the evident traces which the independent statues bear of their derivation from bas-relief: seeing that nearly all of them not only display that union of the limbs with the body which is the characteristic of bas-relief, but have the back of the statue united from head to foot with a block which stands in place of the original wall. Greece repeated the leading stages of this progress. As in Egypt and Assyria, these twin arts were at first united with each other and with their parent, Architecture; and were the aids of Religion and Government. On the friezes of Greek temples, we see coloured bas-reliefs representing sacrifices, battles, processions, games—all in some sort religious. On the pediments we see painted sculptures more or less united with the tympanum, and having for subjects the triumphs of gods or heroes. Even when we come to statues that are definitely separated from the buildings to which they pertain, we still find them coloured; and only in the later periods of Greek civilization, does the differentiation of sculpture from painting appear to have become complete. In Christian art we may clearly trace a parallel re-genesis. All early paintings and sculptures throughout Europe, were religious in subject—represented Christs, crucifixions, virgins, holy families, apostles, saints. They formed integral parts of church architecture, and were among the means of exciting worship: as in Roman Catholic countries they still are. Moreover, the early sculptures of Christ on the cross, of virgins, of saints, were coloured; and it needs but to call to mind the painted madonnas and crucifixes still abundant in continental churches and highways, to perceive the significant fact that painting and sculpture continue in closest connexion with each other, where they continue in closest connexion with their parent. Even when Christian sculpture was pretty clearly differentiated from painting, it was still religious and governmental in its subjects—was used for tombs in churches and statues of kings; while, at the same time, painting, where not purely ecclesiastical, was applied to the decoration of palaces, and besides representing royal personages, was almost wholly devoted to sacred legends. Only in quite recent times have painting and sculpture become entirely secular arts. Only within these few centuries has painting been divided into historical, landscape, marine, architectural, genre, animal, still-life, &c., and sculpture grown heterogeneous in respect of the variety of real and ideal subjects with which it occupies itself.

Strange as it seems then, we find it no less true, that all forms of written language, of painting, and of sculpture, have a common root in the politico-religious decorations of ancient temples and palaces. Little resemblance as they now have, the bust that stands on the console, the landscape that hangs against the wall, and the copy of the *Times* lying upon the table, are remotely akin; not only in nature, but by extraction. The brazen face of the knocker which the postman has just lifted, is related not only to the woodcuts of the *Illustrated London News* which he is delivering, but to the characters of the *billet-doux* which accompanies it. Between the painted window, the prayer-book on which its light falls, and the adjacent monument, there is consanguinity. The effigies on our coins, the signs over shops, the figures that fill every ledger, the coat of arms outside the carriage-panel, and the placards inside the omnibus, are, in common with dolls, blue-books and paper-hangings, lineally descended from the rude sculpture-paintings in which the Egyptians represented the triumphs and worship of their god-kings. Perhaps no example can be given which more vividly illustrates the multiplicity and heterogeneity of the products that in course of time may arise by successive differentiations from a common stock.

Before passing to other classes of facts, it should be observed that the evolution of the homogeneous into the heterogeneous is displayed not only in the separation of Painting and Sculpture from Architecture and from each other, and in the greater variety of subjects they embody; but it is further shown in the structure of each work. A modern picture or statue is of far more heterogeneous nature than an ancient one. An Egyptian sculpture-fresco represents all its figures as on one plane—that is, at the same distance from the eye; and so is less heterogeneous than a painting that represents them as at various distances from the eye. It exhibits all objects as exposed to the same degree of light; and so is less heterogeneous than a painting which exhibits different objects, and different parts of each object, as in different degrees of light. It uses scarcely any but the primary colours, and these in their full intensity; and so is less heterogeneous than a painting which, introducing the primary colours but sparingly, employs an endless variety of intermediate tints, each of heterogeneous composition, and differing from the rest not only in quality but in intensity. Moreover, we see in these earliest works a great uniformity of conception. The same arrangement of figures is perpetually reproduced—the same actions, attitudes, faces, dresses. In Egypt the modes of representation were so fixed that it was sacrilege to introduce a novelty; and indeed it could have been only in consequence of a fixed mode of representation that a system of hieroglyphics became possible. The Assyrian bas-reliefs display parallel characters. Deities, kings, attendants, winged-figures and animals, are severally depicted in like positions, holding like implements, doing like things, and with like expression or non-expression of face. If a palm-grove is introduced, all the trees are of the same height, have the same number of leaves, and are equidistant. When water is imitated, each wave is a counterpart of the rest; and the fish, almost always of one kind, are evenly distributed over the surface. The beards of the kings, the gods, and the winged-figures, are everywhere similar; as are the manes of the lions, and equally so those of the horses. Hair is represented throughout by one form of curl. The king's beard is quite architecturally built up of compound tiers of uniform curls, alternating with twisted tiers placed in a transverse direction, and arranged with perfect regularity; and the terminal tufts of the bulls' tails are represented in exactly the same manner. Without tracing out analogous facts in early Christian art, in which, though less striking, they are still visible, the advance in heterogeneity will be sufficiently manifest on remembering that in the pictures of our own day the composition is endlessly varied; the attitudes, faces, expressions, unlike; the subordinate objects different in size, form, position, texture; and more or less of contrast even in the smallest details. Or, if we compare an Egyptian statue, seated bolt upright on a block, with hands on knees, fingers outspread and parallel, eyes looking straight forward, and the two sides perfectly symmetrical in every particular, with a statue of the advanced Greek or the modern school, which is asymmetrical in respect of the position of the head, the body, the limbs, the arrangement of the hair, dress, appendages, and in its relations to neighbouring objects, we shall see the change from the homogeneous to the heterogeneous clearly manifested.

§ 51. In the co-ordinate origin and gradual differentiation of Poetry, Music, and Dancing, we have another series of illustrations. Rhythm in speech, rhythm in sound, and rhythm in motion, were in the beginning, parts of the same thing; and have only in process of time become separate things. Among various existing barbarous tribes we find them still united. The dances of savages are accompanied by some kind of monotonous chant, the clapping of hands, the striking of rude instruments: there are measured movements, measured words, and measured tones; and the whole ceremony, usually having reference to war or sacrifice, is of governmental character. In the early records of the historic races we similarly find these three forms of metrical action united in religious festivals. In the Hebrew writings we read that the triumphal ode composed by Moses on the defeat of the Egyptians, was sung to an accompaniment of dancing and timbrels. The Israelites danced and sung "at the inauguration of the golden calf. And as it is generally agreed that this representation of the Deity was borrowed from the mysteries of Apis, it is probable that the dancing was copied from that of the Egyptians on those occasions." There was an annual dance in Shiloh on the sacred festival; and David danced before the ark. Again, in Greece the like relation is everywhere seen: the original type being there, as probably in other cases, a simultaneous chanting and mimetic representation of the life and adventures of the god. The Spartan dances were accompanied by hymns and songs; and in general the Greeks had "no festivals or religious assemblies but what were accompanied with songs and dances"—both of them being forms of worship used before altars. Among the Romans, too, there were sacred dances: the Salian and Lupercalian being named as of that kind. And even in Christian countries, as at Limoges in comparatively recent times, the people have danced in the choir in honour of a saint. The incipient separation of these once united arts from each other and from religion, was early visible in Greece. Probably diverging from dances partly religious, partly warlike, as the Corybantian, came the war-dances proper, of which there were various kinds; and from these resulted secular dances. Meanwhile Music and Poetry, though still united, came to have an existence separate from dancing. The aboriginal Greek poems, religious in subject, were not recited but chanted; and though at first the chant of the poet was accompanied by the dance of the chorus, it ultimately grew into independence. Later still, when the poem had been differentiated into epic and lyric—when it became the custom to sing the lyric and recite the epic— poetry proper was born. As during the same period musical instruments were being multiplied, we may presume that music came to have an existence apart from words. And both of them were beginning to assume other forms besides the religious. Facts having like implications might be cited from the histories of later times and peoples; as the practices of our own early minstrels, who sang to the harp heroic narratives versified by themselves to music of their own composition: thus uniting the now separate offices of poet, composer, vocalist, and instrumentalist. But, without further illustration, the common origin and gradual differentiation of Dancing, Poetry, and Music will be sufficiently manifest.

The advance from the homogeneous to the heterogeneous is displayed not only in the separation of these arts from each other and from religion, but also in the multiplied differentiations which each of them afterwards undergoes. Not to dwell upon the numberless kinds of dancing that have, in course of time, come into use; and not to occupy space in detailing the progress of poetry, as seen in the development of the various forms of metre, of rhyme, and of general organization; let us confine our attention to music as a type of the group. As argued by Dr. Burney, and as implied by the customs of still extant barbarous races, the first musical instruments were, without doubt, percussive—sticks, calabashes, tom-toms—and were used simply to mark the time of the dance; and in this constant repetition of the same sound, we see music in its most homogeneous form. The Egyptians had a lyre with three strings. The early lyre of the Greeks had four, constituting their tetrachord. In course of some centuries lyres of seven and eight strings were employed. And, by the expiration of a thousand years, they had advanced to their "great system" of the double octave. Through all which changes there of course arose a greater heterogeneity of melody. Simultaneously there came into use the different modes—Dorian, Ionian, Phrygian, Æolian, and Lydian—answering to our keys: and of these there were ultimately fifteen. As yet, however, there was but little heterogeneity in the time of their music. Instrumental music during this period being merely the accompaniment of vocal music, and vocal music being completely subordinated to words,—the singer being also the poet, chanting his own compositions and making the lengths of his notes agree with the feet of his verses; there unavoidably arose a tiresome uniformity of measure, which, as Dr Burney says, "no resources of melody could disguise." Lacking the complex rhythm obtained by our equal bars and unequal notes, the only rhythm was that produced by the quantity of the syllables, and was of necessity comparatively monotonous. And further, it may be observed that the chant thus resulting, being like recitative, was much less clearly differentiated from ordinary speech than is our modern song. Nevertheless, considering the extended range of notes in use, the variety of modes, the occasional variations of time consequent on changes of metre, and the multiplication of instruments, we see that music had, towards the close of Greek civilization, attained to considerable heterogeneity: not indeed as compared with our music, but as compared with that which preceded it. As yet, however, there existed nothing but melody: harmony was unknown. It was not until Christian church-music had reached some development, that music in parts was evolved; and then it came into existence through a very unobtrusive differentiation. Difficult as it may be to conceive, *à priori*, how the advance from melody to harmony could take place without a sudden leap, it is none the less true that it did so. The circumstance which prepared the way for it, was the employment of two choirs singing alternately the same air. Afterwards it became the practice (very possibly first suggested by a mistake) for the second choir to commence before the first had ceased; thus producing a fugue. With the simple airs then in use, a partially harmonious fugue might not improbably thus result; and a very partially harmonious fugue satisfied the ears of that age, as we know from still preserved examples. The idea having once been given, the composing of airs productive of fugal harmony would naturally grow up; as in some way it *did* grow up out of this alternate choir-singing. And from the fugue to concerted music of two, three, four, and more parts, the transition was easy. Without pointing out in detail the increasing complexity that resulted from introducing notes of various lengths, from the multiplication of keys, from the use of accidentals, from varieties of time, from modulations and so forth, it needs but to contrast music as it is, with music as it was, to see how immense is the increase of heterogeneity. We see this if, looking at music in its *ensemble*, we enumerate its many different genera and species— if we consider the divisions into vocal, instrumental, and mixed; and their subdivisions into music for different voices and different instruments—if we observe the many forms of sacred music, from the simple hymn, the chant, the canon, motet, anthem, &c., up to the oratorio; and the still more numerous forms of secular music, from the ballad up to the serenata, from the instrumental solo up to the symphony. Again, the same truth is seen on comparing any one sample of aboriginal music with a sample of modern music—even an ordinary song for the piano; which we find to be relatively highly heterogeneous, not only in respect of the varieties in the pitch and in the length of the notes, the number of different notes sounding at the same instant in company with the voice, and the variations of strength with which they are sounded and sung, but in respect of the changes of key, the changes of time, the changes of *timbre* of the voice, and the many other modifications of expression. While between the old monotonous dance-chant and a grand opera of our own day, with its endless orchestral complexities and vocal combinations, the contrast in heterogeneity is so extreme that it seems scarcely credible that the one should have been the ancestor of the other.

§ 52. Were they needed, many further illustrations might be cited. Going back to the early time when the deeds of the god-king, chanted and mimetically represented in dances round his altar, were further narrated in picture-writings on the walls of temples and palaces, and so constituted a rude literature, we might trace the development of Literature through phases in which, as in the Hebrew Scriptures it presents in one work, theology, cosmogony, history, biography, civil law, ethics, poetry; through other phases in which, as in the Iliad, the religious, martial, historical, the epic, dramatic, and lyric elements are similarly commingled; down to its present heterogeneous development, in which its divisions and subdivisions are so numerous and varied as to defy complete classification. Or we might track the evolution of Science: beginning with the era in which it was not yet differentiated from Art, and was, in union with Art, the handmaid of Religion; passing through the era in which the sciences were so few and rudimentary, as to be simultaneously cultivated by the same philosophers; and ending with the era in which the genera and species are so numerous that few can enumerate them, and no one can adequately grasp even one genus. Or we might do the like with Architecture, with the Drama, with Dress. But doubtless the reader is already weary of illustrations; and my promise has been amply fulfilled. I believe it has been shown beyond question, that that which the German physiologists have found to be the law of organic development, is the law of all development. The advance from the simple to the complex, through a process of successive differentiations, is seen alike in the earliest changes of the Universe to which we can reason our way back, and in the earliest changes which we can inductively establish; it is seen in the geologic and climatic evolution of the Earth, and of every single organism on its surface; it is seen in the evolution of Humanity, whether contemplated in the civilized individual, or in the aggregation of races; it is seen in the evolution of Society, in respect alike of its political, its religious, and its economical organization; and it is seen in the evolution of all those endless concrete and abstract products of human activity, which constitute the environment of our daily life. From the remotest past which Science can fathom, up to the novelties of yesterday, that in which Evolution essentially consists, is the transformation of the homogeneous into the heterogeneous.

8. The substance of this chapter is nearly identical with the first half of an essay on "Progress: its Law and Cause," which was originally published in the *Westminster Review* for April 1857: only a few unimportant additions and alterations have been made. The succeeding chapter, however, in which the subject is continued, is, with the exception of a fragment embodied in it, wholly new.

9. For detailed proof of these assertions see essay on *Manners and Fashion*.

CHAPTER III
THE LAW OF EVOLUTION, CONTINUED

§ 53. But now, does this generalization express the whole truth? Does it include all the phenomena of Evolution? and does it exclude all other phenomena? A careful consideration of the facts, will show that it does neither.

That there are changes from the less heterogeneous to the more heterogeneous, which do not come within what we call Evolution, is proved in every case of local disease. A portion of the body in which there arises a cancer, or other morbid growth, unquestionably displays a new differentiation. Whether this morbid growth be, or be not, more heterogeneous than the tissues in which it is seated, is not the question. The question is, whether the structure of the organism as a whole, is, or is not, rendered more heterogeneous by the addition of a part unlike every pre-existing part, both in form and composition. And to this question there can be none but an affirmative answer. Again, it might with apparent truth be contended, that the earlier stages of decomposition in a dead body, similarly involve an increase of heterogeneity. Supposing the chemical changes to commence in some parts of the body earlier than in other parts, as they commonly do; and to affect different tissues in different, ways, as they must; it seems to be a necessary admission that the entire body, made up of undecomposed parts and parts decomposed in different ways and degrees, has become more heterogeneous than it was. Though greater homogeneity will be the eventual result, the immediate result is the opposite. And yet this immediate result is certainly not evolution. But perhaps of all illustrations the least debatable are those furnished by social disorders and disasters. When in any nation there occurs a rebellion, which, while leaving some provinces undisturbed, developes itself here in secret societies, there in public demonstrations, and elsewhere in actual appeal to arms, leading probably to conflict and bloodshed; it must be admitted that the society, regarded as a whole, has so been rendered more heterogeneous. Or when a dearth causes commercial panic with its entailed bankruptcies, closed factories, discharged operatives, political agitations, food riots, incendiarisms; it is manifest that as, throughout the rest of society, there still exists the ordinary organization displaying the usual phenomena, these new phenomena must be regarded as adding to the complexity previously existing. Nevertheless, it is clear that such changes so far from constituting a further stage of evolution, are steps towards dissolution.

There is good reason to think then, that the definition arrived at in the last chapter, is an imperfect one. We may suspect, not that the process of evolution is different from the process there described; but that the description did not contain all that it should. The changes above instanced as coming within the formula as it now stands, are so obviously different from the rest, that the inclusion of them implies some oversight—some distinction hitherto overlooked. Such further distinction we shall find really exists.

§54. At the same time that all evolution is a change from the homogeneous to the heterogeneous, it is also a change from the indefinite to the definite. As well as an advance from simplicity to complexity, there is an advance from confusion to order—from undetermined arrangement to determined arrangement. In the process of development, no matter what sphere it is displayed in, there is not only a gradual multiplication of unlike parts; but there is a gradual increase in the distinctness with which these parts are marked off from each other. And so is that increase of heterogeneity which characterizes Evolution, distinguished from that increase of heterogeneity which does not. For proof of this, it needs only to reconsider the instances given above. The structural changes constituting a disease, have no such definiteness, either in locality, extent, or outline, as the structural changes constituting development. Though certain morbid growths arise much more commonly in some parts of the body than in others (as warts on the hands, cancer on the breasts, tubercle in the lungs), yet they are not confined to these parts; nor, when found on them, are they anything like so precise in their relative positions as are the normal parts around them. In size, again, they are extremely variable—they bear no such constant proportion to the body as organs do. Their forms, too, are far less specific than organic forms. And they are extremely irregular or confused in their internal structures. That is to say, they are in all respects comparatively indefinite. The like peculiarity may be traced in decomposition. That state of total indefiniteness to which a dead body is finally reduced, is a state towards which the putrefactive changes have tended from their commencement. Each step in the destruction of the organic compounds, is accompanied by a blurring of the minute structure—diminishes its distinctness. From the portions that have undergone most decomposition, there is a gradual transition to the less decomposed portions. And step by step the lines of organization, once so precise, disappear. Similarly with social changes of an abnormal kind. A political outbreak rising finally to a rebellion, tends from the very first to obliterate the specializations, governmental and industrial, which previously existed. The disaffection which originates such an outbreak, itself implies a loosening of those ties by which the citizens are bound up into distinct classes and sub-classes. Agitation, growing into revolutionary meetings, shows us a decided tendency towards the fusion of ranks that are usually separated. Acts of open insubordination exhibit a breaking through of those definite limits to individual conduct which were previously observed; and a disappearance of the lines previously existing between those in authority and those beneath them. At the same time, by the arrest of trade, artizans and others lose their occupations; and in so ceasing to be functionally distinguished, become fused into a mass from which the demarcations in great part vanish. And when at last there comes positive insurrection, all magisterial and official powers, all class distinctions, and all industrial differences, at once cease: organized society lapses into an unorganized aggregation of social units. How the like holds true of such social disasters as are entailed by famine, needs not be pointed out. On calling to mind that in cases of this kind the changes are from order towards disorder, it will at once be seen that like the foregoing they are changes from definite arrangements to indefinite arrangements.

Thus then is that increase of heterogeneity which constitutes Evolution, distinguished from that increase of heterogeneity which does not do so. Though in disease and death, individual or social, the earliest modifications may be construed as additions to the heterogeneity previously existing; yet they cannot be construed as additions to the definiteness previously existing. They begin from the very outset to destroy this definiteness; and so, gradually produce a heterogeneity that is indeterminate instead of determinate. Just in the same way that a city, already multiform in its variously arranged structures of various architecture, may be made more multiform by an earthquake, which leaves part of it standing and overthrows other parts in different ways and degrees, and yet is at the same time reduced from definite arrangement to indefinite arrangement; so may organized bodies be made for a time more multiform by changes which are nevertheless disorganizing changes. And in the one case as in the other, it is the absence of definiteness which distinguishes the multiformity of regression from the multiformity of progression.

If the advance from the indefinite to the definite is an essential characteristic of Evolution, we shall of course find it everywhere displayed; as in the last Chapter we found the advance from the homogeneous to the heterogeneous. With a view of showing that it is so, let us now briefly reconsider the same several classes of facts.

§ 55. Beginning as before with a hypothetical illustration, we have to note that each further stage in the evolution of the Solar System, supposing it to have originated from diffused matter, was an advance towards more definite forms, and times, and forces. At first irregular in shape and with indistinct margins, the attenuated substance, as it concentrated and acquired a rotatory motion, must have assumed the shape of an oblate spheroid; which, with every increase of density, became more specific in general outline, and had its surface more sharply marked off from the surrounding void. At the same time, the constituent portions of nebulous matter, instead of independently moving towards their common centre of gravity from all points, and tending to revolve round it in various planes, as they would at first do, must have had these planes more and more merged into a single plane; and this plane must have gained greater precision as the concentration progressed. To which add that in the gradual establishment of a common and determinate angular velocity, instead of the various and conflicting angular velocities of different parts, we have a further change of like nature. According to the hypothesis, change from indistinct characteristics to distinct ones, was repeated in the evolution of each planet and satellite; and may in them be traced to a much greater extent. A gaseous spheroid is less definitely marked off from the space around it than a fluid spheroid, since it is subject to larger and more rapid undulations of surface, and to much greater distortions of general form; and similarly a fluid spheroid, covered as it must be with waves of various magnitudes, is less definite than a solid spheroid. Nor is it only in greater fixity of surface that a planet in its last stage, is distinguished from a planet in its earlier stages. Its general form, too, is more precise. The sphere, to which in the end it very closely approximates, is a perfectly specific figure; while the spheroid, under which figure it previously existed, being infinitely variable in oblateness, is an imperfectly specific figure. And further, a planet having an axis inclined to the plane of its orbit, must, while its form is very oblate, have its plane of rotation greatly disturbed by the attraction of external bodies; whereas its approach to a spherical form, involving a less extreme precessional motion, implies less marked variations in the direction of its axis. Nor is it only in respect of space-relations that the Solar System in general and in detail has become more precise. The like is true of time-relations. During the process of concentration the various portions of the nebulous mass must not only differ more or less from each other in their angular velocities, but each of them must gradually change the period in which it moves round the general axis. In every detached ring however, and in the resulting planet, this progressive alteration ceases: there results a determinate period of revolution. And similarly the time of axial rotation, which, during the formation of each planet, is continually diminishing, becomes at last practically fixed: as in the case of the Earth, whose day is not a second less than it was 2000 years ago. It is scarcely needful to point out that the force-relations have simultaneously become more and more settled. The exact calculations of physical astronomy, show us how definite these force-relations now are; while the great indefiniteness which once characterized them, is implied in the extreme difficulty, if not impossibility, of subjecting the nebular hypothesis to mathematical treatment.

From that originally molten state of the Earth inferable from established geological data—a state in harmony with the nebular hypothesis but inexplicable on any other—the transition to its existing state has been through stages in which the characters became more determinate. Besides being, as above pointed out, comparatively unstable in surface and contour, a fluid spheroid is less definite than a solid spheroid in having no fixed distribution of parts. Currents of molten matter, though kept to certain general circuits by the conditions of equilibrium, cannot in the absence of solid boundaries be precise or permanent in direction: all parts must be in motion with respect to other parts. But a solidification of the surface, even though but partial, is manifestly a step towards the establishment of definite relations of position. In a thin crust however, frequently ruptured as it must be by disturbing forces, and moved by every tidal undulation, such fixity of relative position can be but temporary. Only as the crust slowly increases in thickness, can there arise distinct and settled geographical relations. Observe too that when, on a crust that has cooled to the requisite degree, there begins to precipitate the water floating above as vapour, the water which is precipitated cannot maintain any definiteness either of state or place. Falling on a surface not thick enough to preserve anything beyond slight variations of level, it must form small shallow deposits over areas sufficiently cool to permit condensation; which areas must not only pass insensibly into others that are too hot for this, but must themselves from time to time be so raised in temperature as to drive off the water lying on them. With progressive refrigeration, however,—with an increasing thickness of crust, a consequent formation of larger elevations and depressions, and the condensation of more atmospheric water, there comes an arrangement of parts that is comparatively fixed in both time and space; and the definiteness of state and position increases, until there results such a distribution of continents and oceans as we now see—a distribution that is not only topographically precise, but also in its cliff-marked coast-lines presents a more definite division of land from water than could have existed during the period when islands of low elevation had shelving beaches up which the tide ebbed and flowed to great distances. Respecting the characteristics technically classed as geological, we may draw parallel inferences. While the Earth's crust was thin, mountain-chains were impossibilities: there could not have been long and well-defined axes of elevation, with distinct water-sheds and areas of drainage. Moreover, from small islands admitting of but small rivers, and tidal streams both feeble and narrow, there would result no clearly-marked sedimentary strata. Confused and varying masses of detritus, such as those now found at the mouths of brooks, must have been the prevailing formations. And these could give place to distinct strata, only as there arose continents and oceans, with their great rivers, long coast-lines, and wide-spreading marine currents. How there must simultaneously have resulted more definite meteorological characters, need not be pointed out in detail. That differences of climates and seasons must have grown relatively decided as the heat of the Sun became distinguishable from the proper heat of the Earth; that the establishment through this cause of comparatively constant atmospheric currents, must have similarly produced more specific conditions in each locality; and that these effects must have been aided by increasing permanence in the distribution of land and sea and of ocean currents; are conclusions which are sufficiently obvious.

Let us turn now to the evidence furnished by organic bodies. In place of deductive illustrations like the foregoing, we shall here find numerous illustrations which, as being inductively established, are less open to criticism. The process of mammalian development, for example, will supply us with numerous proofs ready-described by embryologists. The first change which the ovum of a mammal undergoes, after continued segmentation has reduced its yelk to a mulberry-like mass, is the appearance of a greater definiteness in the peripheral cells of this mass: each of which acquires a distinct enveloping membrane. These peripheral cells, vaguely distinguished from the internal ones both by their greater completeness and by their minuter subdivision, coalesce to form the blastoderm or germinal membrane. One portion of the blastoderm presently becomes contrasted with the rest, through the accumulation of cells still more subdivided, which, together, form an opaque roundish spot. This *area germinativa*, as it is called, is not sharply delineated, but shades off gradually into the surrounding parts of the blastoderm; and the *area pellucida*, subsequently formed in the midst of this germinal area, is similarly without any precise margin. The "primitive trace," which makes its appearance in the centre of the *area pellucida*, and is the rudiment of that vertebrate axis which is to be the fundamental characteristic of the mature animal, is shown by its name to be at first indefinite—a mere trace. Beginning as a shallow groove, this becomes slowly more pronounced: its sides grow higher, their summits overlap, and at last unite; and so the indefinite groove passes into a definite tube, forming the vertebral canal. In this vertebral canal the leading divisions of the brain are at first discernible only as slight bulgings; while the vertebræ commence as indistinct modifications of the tissue bounding the canal. Simultaneously, the outer portion of the blastoderm has been undergoing separation from the inner portion: there has been a division into the serous and mucous layers—a division at the outset indistinct, and traceable only about the germinal area, but which insensibly spreads throughout nearly the whole germinal membrane, and becomes definite. From the mucous layer, the development of the alimentary canal proceeds as that of the vertebral canal does from the serous layer. Originally a simple channel along the under surface of the embryonic mass, the intestine is rendered step by step more distinct by the bending down, on each side, of ridges which finally join to form a tube—the permanent absorbing surface is by degrees clearly cut off from that temporary absorbing surface of which it was at first a part like all the rest. And in an analogous manner the entire embryo, which at first lies outspread upon the surface of the yelk-sack, gradually rises up from it, and, by the infolding of its ventral surface, becomes a separate mass, connected with the yelk-sack only by a narrow duct. These changes through which the general structure of the embryo is marked out with slowly-increasing precision, are paralleled in the evolution of each organ. The heart is at first a mere aggregation of cells, of which the inner liquify to form the cavity, while the outer are transformed into the walls; and when thus sketched out, the heart is indefinite not only as being unlined by limiting membrane, but also as being but vaguely distinguishable from the great blood-vessels: of which it is little more than a dilatation. By and by the receiving portion of the cavity becomes distinct from the propelling portion. Afterwards there begins to be formed across the ventricle, a septum, which, however, is some time before it completely shuts off the two halves from each other; while the later-formed septum of the auricle remains incomplete during the whole of fœtal life. Again, the liver commences as a multiplication of certain cells in the wall of the intestine. The thickening produced by this multiplication "increases so as to form a projection upon the exterior of the canal;" and at the same time that the organ grows and becomes distinct from the intestine, the channels which permeate it are transformed into ducts having clearly-marked walls. Similarly, by the increase of certain cells of the external coat of the alimentary canal at its upper portion, are produced buds from which the lungs are developed; and these, in their general outlines and detailed structure, acquire distinctness step by step. Changes of this order continue long after birth; and, in the human being, are some of them not completed till middle life. During youth, most of the articular surfaces of the bones remain rough and fissured—the calcareous deposit ending irregularly in the surrounding cartilage. But between puberty and the age of thirty, the articular surfaces are finished off by the addition of smooth, hard, sharply-cut "epiphyses." Thus we may say that during Evolution, an increase of definiteness continues long after there ceases to be any appreciable increase of heterogeneity. And, indeed, there is reason to think that those structural modifications which take place after maturity, ending in old age and death, are modifications of this nature; since they result in a growing rigidity of structure, a consequent restriction of movement and of functional pliability, a gradual narrowing of the limits within which the vital processes go on, ending at length in an organic adjustment too precise—too narrow in its margin of possible variation to permit the requisite adaptation to external changes of condition.

To demonstrate that the Earth's Flora and Fauna, regarded either as wholes or in their separate species, have progressed in definiteness, is of course no more possible than it was to demonstrate that they have progressed in heterogeneity: lack of facts being an obstacle to the one conclusion as to the other. If, however, we allow ourselves to reason from the hypothesis, now daily rendered more probable, that every species of organic form up to the most complex, has arisen out of the simplest through the accumulation of modifications upon modifications, just as every individual organic form arises; we shall see that in such case there must have been a progress from the indeterminate to the determinate, both in the particular forms and in the groups of forms. We may set out with the significant fact that many of the lowest living organisms (which are analogous in structure to the germs of all higher ones) are so indefinite in character that it is difficult, if not impossible, to decide whether they are plants or animals. Respecting sundry of them there are unsettled disputes between zoologists and botanists; and it has even been proposed to group them into a separate kingdom, forming a common basis to the animal and vegetal kingdoms. Note next that among the *Protozoa*, extreme indefiniteness of shape is very general. In the shell-less Rhizopods and their allies, not only is the form so irregular as to admit of no description, but it is neither alike in any two individuals nor in the same individual at successive moments. By the aggregation of such creatures, are produced, among other indefinite bodies, the sponges—bodies that are indefinite in size, in contour, in internal arrangement, and in the absence of an external limiting membrane. As further showing the relatively indeterminate character of the simplest organisms, it may be mentioned that their structures vary very greatly with surrounding conditions: so much so that, among the *Protozoa* and *Protophyta*, many forms which were once classed as distinct species, and even as distinct genera, are found to be merely varieties of one species. If now we call to mind how precise in their attributes are the highest organisms—how sharply cut their outlines, how invariable their proportions, and how comparatively constant their structures under changed conditions, we cannot deny that greater definiteness is one of their characteristics; and that if they have been evolved out of lower organisms, an increase of definiteness has been an accompaniment of their evolution. That in course of time, species have become more sharply marked off from other species, genera from genera, and orders from orders, is a conclusion not admitting of a more positive establishment than the foregoing; and must, indeed, stand or fall with it. If, however, species and genera and orders have resulted from the process of "natural selection," then, as Mr. Darwin shows, there must have been a tendency to divergence, causing the contrasts between groups to become more and more pronounced. By the disappearance of intermediate forms, less fitted for special spheres of existence than the extreme forms they connected, the differences between the extreme forms must be rendered more decided; and so, from indistinct and unstable varieties, must slowly be produced distinct and stable species. Of which inference it may be remarked, not only that it follows from a process to which the organic creation is of necessity ever subject, but also that it is in harmony with what we know respecting races of men and races of domestic animals.

Evidence that in the course of psychial development, there is a change from the vague to the distinct, may be seen in every nursery. The confusion of the infant's perceptions is shown by its inability to distinguish persons. The dimness of its ideas of direction and distance, may be inferred from the ill-guided movements of its hands, and from its endeavours to grasp objects far out of reach. Only by degrees does the sense of equilibrium, needful for safe standing and moving, gain the requisite precision. Through the insensible steps that end in comprehensible speech, we may trace an increase in the accuracy with which sounds are discriminated and in the nicety with which they are imitated. And similarly during education, the change is towards the establishment of internal relations more perfectly corresponding to external ones—to exactness in calculations, to a better representation of objects drawn, to a more correct spelling, to a completer conformity to the rules of speech, to clearer ideas respecting the affairs of life. How in the further progress to maturity the law still holds, needs not here be pointed out; more especially as it will presently be shown in treating of the evolution of intelligence during the advance of civilization. The only further fact calling for remark, is, that this increase of mental definiteness is, in some ways, manifested even during the advance from maturity to old age. The habits of life grow more and more fixed; the character becomes less capable of change; the quantity of knowledge previously acquired ceases to have its limits alterable by additions; and the opinions upon every point admit of no modification.

Still more manifestly do the successive phases through which societies pass, display the progress from indeterminate arrangement to determinate arrangement. A wandering tribe of savages, as being fixed neither in its locality nor in the relative positions of its parts, is far less definite than a nation, covering a territory clearly marked out, and formed of individuals grouped together in towns and villages. In such a tribe the social relations are similarly confused and unsettled. Political authority is neither well established nor precise. Distinctions of rank are neither clearly marked nor impassable. "Medicine-men" and "rain-makers" form a class by no means as distinct from the rest of the community as eventually becomes the priesthood they foreshadow. And save in the different occupations of men and women, there are no complete industrial divisions. Only in tribes of considerable size, which have enslaved other tribes, is the economical differentiation decided. Any one of these primitive societies however that developes, becomes step by step more specific. Increasing in size, consequently ceasing to be so nomadic, and restricted in its range by neighbouring tribes, it acquires, after prolonged border warfare, a more settled territorial boundary. The distinction between the royal race and the people, grows so extreme as to amount in the popular apprehension to a difference of nature. The warrior-class attains a perfect separation from classes devoted to the cultivation of the soil or other occupations regarded as servile. And there arises a priesthood that is defined in its rank, its functions, its privileges. This sharpness of definition, growing both greater and more variously exemplified as societies advance to maturity, is extremest in those that have reached their full development or are declining. Of ancient Egypt we read that its social divisions were strongly-marked and its customs rigid. Recent investigations make it more than ever clear, that among the Assyrians and surrounding peoples, not only were the laws unalterable, but even the minor habits, down to those of domestic routine, possessed a sacredness which insured their permanence. In India at the present day, the unchangeable distinctions of caste, not less than the constancy in modes of dress, industrial processes, and religious observances, show us how fixed are the arrangements where the antiquity is great. Nor does China with its long-settled political organization, its elaborate and precise conventions, and its unprogressive literature, fail to exemplify the same truth. The successive phases of our own and neighbouring societies, furnish facts somewhat different in kind but similar in meaning. After our leading class-divisions had become tolerably well-established, it was long before they acquired their full precision. Originally, monarchical authority was more baronial, and baronial authority more monarchical, than they afterwards became. Between modern priests and the priests of old times, who while officially teachers of religion were also warriors, judges, architects, there is a marked difference in definiteness of function. And among the people engaged in productive occupations, the like contrast would be found to hold: the industrial office has become more distinct from the military; and its various divisions from each other. A history of our constitution, reminding us how, after prolonged struggles, the powers of King, Lords, and Commons, have been gradually settled, would clearly exhibit analogous changes. Countless facts bearing the like construction would meet us, were we to trace the development of legislation: in the successive stages of which, we should find statutes made more precise in their provisions—more specific in their applications to particular cases. Even at the present time we see that each new law, beginning as a vague proposition, is, in the course of enactment, elaborated into specific clauses; and further that only after its interpretation has been established by judges' decisions in courts of justice, does it reach its final definiteness. From the history of minor institutions like evidence may be gathered. Religious, charitable, literary, and all other societies, beginning with ends and methods roughly sketched out and easily modifiable, show us how, by the accumulation of rules and precedents, the purposes become more distinct and the modes of action more restricted; until at last death often results from a fixity which admits of no adaptation to new conditions. Should it be objected that among civilized nations there are examples of decreasing definiteness, (instance the breaking down of limits between ranks,) the reply is, that such apparent exceptions are the accompaniments of a social metamorphosis—a change from the military or predatory type of social structure, to the industrial or mercantile type, during which the old lines of organization are disappearing and the new ones becoming more marked.

That all organized results of social action, pass in the course of civilization through parallel phases, is demonstrable. Being, as they are, objective products of subjective processes, they must display corresponding changes; and that they do this, the cases of Language, of Science, of Art, clearly prove.

If we strike out from our sentences everything but nouns and verbs, we shall perceive how extremely vague is the expression of ideas in undeveloped tongues. When we note how each inflection of a verb or addition by which the case of a noun is marked, serves to limit the conditions of action or of existence, we see that these constituents of speech enable men more precisely to communicate their thoughts. That the application of an adjective to a noun or an adverb to a verb, narrows the class of things or changes indicated, implies that these additional words serve further to define the meaning. And similarly with other parts of speech. The like effect results from the multiplication of words of each order. When the names for objects, and acts, and qualities, are but few, the range of each is proportionately wide, and its meaning therefore unspecific. The similes and metaphors so abundantly used by aboriginal races, are simply vehicles for indirectly and imperfectly conveying ideas, which lack of words disables them from conveying directly and perfectly. In contrasting these figurative expressions, interpretable in various senses, with the expressions which we should use in place of them, the increase of exactness which wealth of language gives, is rendered very obvious. Or to take a case from ordinary life, if we compare the speech of the peasant, who, out of his limited vocabulary, can describe the contents of the bottle he carries, only as "doctor's-stuff" which he has got for his "sick" wife, with the speech of the physician, who tells those educated like himself the particular composition of the medicine, and the particular disorder for which he has prescribed it; we have vividly brought home to us, the precision which language gains by the multiplication of terms. Again, in the course of its evolution, each tongue acquires a further accuracy through processes which fix the meaning of each word. Intellectual intercourse tends gradually to diminish laxity of expression. By and by dictionaries give definitions. And eventually, among the most cultivated, indefiniteness is not tolerated, either in the terms used or in their grammatical combinations. Once more, languages considered as wholes, become gradually more distinct from each other, and from their common parent: as witness in early times the divergence from the same root of two languages so unlike as Greek and Latin, and in later times the development of three Latin dialects into Italian, French, and Spanish.

In his "History of the Inductive Sciences," Dr. Whewell says that the Greeks failed in physical philosophy because their "ideas were not distinct, and appropriate to the facts." I do not quote this remark for its luminousness; since it would be equally proper to ascribe the indistinctness and inappropriateness of their ideas to the imperfection of their physical philosophy; but I quote it because it serves as good evidence of the indefiniteness of primitive science. The same work and its fellow on "The Philosophy of the Inductive Sciences," supply other evidences equally good, because equally independent of any such hypothesis as is here to be established. Respecting mathematics we have the fact that geometrical theorems grew out of empirical methods; and that these theorems, at first isolated, did not acquire the clearness which complete demonstration gives, until they were arranged by Euclid into a series of dependent propositions. At a later period the same general truth was exemplified in the progress from the "method of exhaustions" and the "method of indivisibles" to the "method of limits;" which is the central idea of the infinitesimal calculus. In early mechanics, too, may be traced a dim perception that action and re-action are equal and opposite; though for ages after, this truth remained unformulated. And similarly, the property of inertia, though not distinctly comprehended until Kepler lived, was vaguely recognized long previously. "The conception of statical force," "was never presented in a distinct form till the works of Archimedes appeared;" and "the conception of accelerating force was confused, in the mind of Kepler and his contemporaries, and did not become clear enough for purposes of sound scientific reasoning before the succeeding century." To which specific assertions may be added the general remark, that "terms which originally, and before the laws of motion were fully known, were used in a very vague and fluctuating sense, were afterwards limited and rendered precise." When we turn from abstract scientific conceptions to the concrete previsions of science, of which astronomy furnishes us with numerous examples, the like contrast is visible. The times at which celestial phenomena will occur, have been predicted with ever-increasing accuracy: errors once amounting to days, have been reduced down to seconds. The correspondence between the real and supposed forms of orbits, has been growing gradually more precise. Originally thought circular, then epicyclical, then elliptical, orbits are now ascertained to be curves which always deviate more or less from perfect ellipses, and which are ever undergoing change. But the general advance of Science in definiteness, is best shown by the contrast between its qualitative stage, and its quantitative stage. At first, the facts ascertained were, that between such and such phenomena some connexion existed—that the appearances *a* and *b* always occurred together or in succession; but it was neither known what was the nature of the relation between *a* and *b*, nor how much of *a* accompanied so much of *b*. The development of Science has in part been the reduction of these vague connexions to distinct ones. Most relations have been determined as belonging to the classes mechanical, chemical, thermal, electric, magnetic, &c.; and we have learnt to infer the amounts of the antecedents and consequents from each other with an exactness that becomes ever greater. Were there space to state them, illustrations of this truth might be cited from all departments of physics; but it must suffice here to instance the general progress of chemistry. Besides the conspicuous fact that we have positively ascertained the constituent elements of an immense number of compounds which our ancestors could not analyze, and of a far greater number which they never even saw, there is the still more conspicuous fact that the combining equivalents of these elements are accurately calculated. The beginnings of a like advance from qualitative to quantitative prevision, may be traced even in some of the higher sciences. Physiology shows it in the weighing and measuring of organic products, and of the materials consumed. By Pathology it is displayed in the use of the statistical method of determining the sources of diseases, and the effects of treatment. In Zoology and Botany, the numerical comparisons of Floras and Faunas, leading to specific conclusions respecting their sources and distributions, illustrate it. And in Sociology, questionable as are the conclusions usually drawn from the classified sum-totals of the census, from Board-of-Trade tables, and from criminal returns, it must be admitted that these imply a progress towards more accurate conceptions of social phenomena. That an essential characteristic of advancing Science is increase in definiteness, appears indeed almost a truism, when we remember that Science may be described as definite knowledge, in contradistinction to that indefinite knowledge possessed by the uncultured. And if, as we cannot question, Science has, in the slow course of ages, been evolved out of this indefinite knowledge of the uncultured; then, the gradual acquirement of that great definiteness which now distinguishes it, must have been a leading trait in its evolution.

The Arts, industrial and æsthetic, furnish illustrations perhaps still more striking. Flint implements of the kind recently found in certain of the later geologic deposits—implements so rude that some have held them to be of natural rather than of artificial origin—show the extreme want of precision in men's first handyworks. Though a great advance on these is seen in the tools and weapons of existing savage tribes, yet an inexactness in forms and fittings, more than anything else distinguishes such tools and weapons from those of civilized races. In a less degree, the productions of semi-barbarous nations are characterized by like defects. A Chinese junk with all its contained furniture and appliances, nowhere presents a perfectly straight line, a uniform curve, or a true surface. Nor do the utensils and machines of our ancestors fail to exhibit a similar inferiority to our own. An antique chair, an old fireplace, a lock of the last century, or almost any article of household use that has been preserved for a few generations, will prove by contrast how greatly the industrial products of our time excel those of the past in their accuracy. Since planing machines have been invented, it has become possible to produce absolutely straight lines, and surfaces so truly level as to be air-tight when applied to each other. While in the dividing-engine of Troughton, in the micrometer of Whitworth, and in microscopes that show fifty thousand divisions to the inch, we have an exactness as far exceeding that reached in the works of our great-grandfathers, as theirs exceeded that of the aboriginal celt-makers. In the Fine Arts there has been a parallel process. From the rudely carved and painted idols of savages, through the early sculptures characterized by limbs having no muscular detail, wooden-looking drapery, and faces devoid of individuality, up to the later statues of the Greeks or some of those now produced, the increased accuracy of representation is conspicuous. Compare the mural paintings of the Egyptians with the paintings of medieval Europe, or these with modern paintings, and the more precise rendering of the appearances of objects is manifest. So too is it with the delineations of fiction and the drama. In the marvellous tales current among Eastern nations, in the romantic legends of feudal Europe, as well as in the mystery-plays and those immediately succeeding them, we see great want of correspondence to the realities of life; not only in the predominance of supernatural events and extremely improbable coincidences, but also in the vaguely-indicated personages, who are nothing more than embodiments of virtue and vice in general, or at best of particular virtues and vices. Through transitions that need not be specified, there has been a progressive diminution, in both fiction and the drama, of whatever is unnatural—whatever does not answer to real life. And now, novels and plays are applauded in proportion to the fidelity with which they exhibit individual characters with their motives and consequent actions; improbabilities, like the impossibilities which preceded them, are disallowed; and there is even an incipient abandonment of those elaborate plots which the realities of life rarely if ever furnish.

Were it needful, it would be easy to accumulate evidences of various other kinds. The progress from myths and legends, extreme in their misrepresentations, to a history that has slowly become, and is still becoming, more accurate; the establishment of settled systematic methods of doing things, instead of the indeterminate ways at first pursued; and the great increase in the number of points on which conflicting opinion has settled down into exact knowledge; might severally be used further to exemplify the general truth enunciated. The basis of induction is, however, already sufficiently wide. Proof that all Evolution is from the indefinite to the definite, we find to be not less abundant than proof that all Evolution is from the homogeneous to the heterogeneous. The one kind of change is co-extensive with the other—is equally with it exhibited throughout Nature.

§ 56. To form a complete conception of Evolution, we have to contemplate it under yet another aspect. This advance from the indefinite to the definite, is obviously not primary but secondary—is an incidental result attendant on the finishing of certain changes. The transformation of a whole that was originally uniform, into a combination of multiform parts, implies a progressive separation. While this is going on there must be indistinctness. Only as each separated division draws into its general mass those diffused peripheral portions which are at first imperfectly disunited from the peripheral portions of neighbouring divisions, can it acquire anything like a precise outline. And it cannot become perfectly definite until its units are aggregated into a compact whole. That is to say, the acquirement of definiteness is simply a concomitant of complete union of the elements constituting each component division. Thus, Evolution is characterized not only by a continuous multiplication of parts, but also by a growing oneness in each part. And while an advance in heterogeneity results from progressive differentiation, an advance in definiteness results from progressive integration. The two changes are simultaneous; or are rather opposite aspects of the same change. This change, however, cannot be rightly comprehended without looking at both its sides. Let us then once more consider Evolution under its several manifestations; for the purpose of noting how it is throughout a process of integration.

The illustrations furnished by the Solar System, supposing it to have had a nebular origin, are so obvious as scarcely to need indicating. That as a whole, it underwent a gradual concentration while assuming its present distribution of parts; and that there subsequently took place a like concentration of the matter forming each planet and satellite, is the leading feature of the hypothesis. The process of integration is here seen in its simplest and most decided form.

Geologic evolution, if we trace it up from that molten state of the Earth's substance which we are obliged to postulate, supplies us with more varied facts of like meaning. The advance from a thin crust, at first everywhere fissured and moveable, to a crust so solid and thick as to be but now and then very partially dislocated by disturbing forces, exemplifies the unifying process; as does likewise the advance from a surface covered with small patches of land and water, to one divided into continents and oceans—an advance also resulting from the Earth's gradual solidification. Moreover, the collection of detritus into strata of great extent, and the union of such strata into extensive "systems," becomes possible only as surfaces of land and water become wide, and subsidences great, in both area and depth; whence it follows that integrations of this order must have grown more pronounced as the Earth's crust thickened. Different and simpler instances of the process through which mixed materials are separated, and the kindred units aggregated into masses, are exhibited in the detailed structure of the Earth. The phenomena of crystallization may be cited *en masse*, as showing how the unifications of similar elements take place wherever the conditions permit. Not only do we see this where there is little or no hindrance to the approach of the particles, as in the cases of crystals formed from solutions, or by sublimation; but it is also seen where there are great obstacles to their approach. The flints and the nodules of iron pyrites that are found in chalk, as well as the silicious concretions which occasionally occur in limestone, can be interpreted only as aggregations of atoms of silex or sulphuret of iron, originally diffused almost uniformly through the deposit, but gradually collected round certain centres, notwithstanding the solid or semi-solid state of the surrounding matter. Iron-stone as it ordinarily occurs, presents a similar phenomenon to be similarly explained; and what is called bog iron-ore supplies the conditions and the result in still more obvious correlation.

During the evolution of an organism, there occurs, as every physiologist knows, not only separation of parts, but coalescence of parts. In the mammalian embryo, the heart, at first a long pulsating blood-vessel, by and by twists upon itself and becomes integrated. The layer of bile-cells constituting the rudimentary liver, do not simply become different from the wall of the intestine in which they at first lie; but they simultaneously diverge from it and consolidate into an organ. The anterior segments of the cerebro-spinal axis, which are at first continuous with the rest, and distinguished only by their larger size, undergo a gradual union; and at the same time the resulting head consolidates into a mass clearly marked off from the rest of the vertebral column. The like process, variously exemplified in other organs, is meanwhile exhibited by the body as a whole; which becomes integrated, somewhat in the same way that the contents of an outspread handkerchief become integrated when its edges are drawn in and fastened to make a bundle. Analogous changes go on long after birth, and continue even up to old age. In the human being that gradual solidification of the bony framework, which, during childhood, is seen in the coalescence of portions of the same bone ossified from different centres, is afterwards seen in the coalescence of bones that were originally distinct. The appendages of the vertebræ unite with the vertebral centres to which they belong—a change not completed until towards thirty. At the same time the epiphyses, formed separately from the main bodies of their respective bones, have their cartilaginous connexions turned into osseous ones—are fused to the masses beneath them. The component vertebræ of the sacrum, which remain separate till about the sixteenth year, then begin to unite; and in ten or a dozen years more their union is complete. Still later occurs the coalescence of the coccygeal vertebræ; and there are some other bony unions which are not completed until advanced age. To which add that the increase of density and toughness, going on throughout the tissues in general during life, may be regarded as the formation of a more highly integrated substance. The species of change thus illustrated under its several aspects in the unfolding of the human body, may be traced in all animals. That mode of it which consists in the union of homogeneous parts originally separate, has been described by Milne-Edwards and others, as exhibited in various of the invertebrata; though it does not seem to have been included by them as an essential peculiarity in the process of organic development. We shall, however, be led strongly to suspect that progressive integration should form part of the definition of this process, when we find it displayed not only in tracing up the stages passed through by every embryo, but also in ascending from the lower living creatures to the higher. And here, as in the evolution of individual organisms, it goes on both longitudinally and transversely: under which different forms we may indeed most conveniently consider it. Of *longitudinal integration*, the sub-kingdom *Annulosa* supplies abundant examples. Its lower members, such as worms and myriapods, are mostly characterized by the great number of segments composing them: reaching in some cases to several hundreds. But in the higher divisions—crustaceans, insects, and spiders—we find this number reduced down to twenty-two, thirteen, or even fewer; while, accompanying the reduction, there is a shortening or integration of the whole body, reaching its extreme in the crab and the spider. The significance of these contrasts, as bearing upon the general doctrine of Evolution, will be seen when it is pointed out that they are parallel to those which arise during the development of individual *Annulosa*. In the lobster, the head and thorax form one compact box, made by the union of a number of segments which in the embryo were separable. Similarly, the butterfly shows us segments so much more closely united than they were in the caterpillar, as to be, some of them, no longer distinguishable from each other. The *Vertebrata* again, throughout their successively higher classes, furnish like instances of longitudinal union. In most fishes, and in reptiles that have no limbs, the only segments of the spinal column that coalesce, are those forming the skull. In most mammals and in birds, a variable number of vertebræ become fused together to form the sacrum; and in the higher quadrumana and man, the caudal vertebræ also lose their separate individualities in a single *os coccygis*. That which we may distinguish as *transverse integration*, is well illustrated among the *Annulosa* in the development of the nervous system. Leaving out those most degraded forms which do not present distinct ganglia, it is to be observed that the lower annulose animals, in common with the larvæ of the higher, are severally characterized by a double chain of ganglia running from end to end of the body; while in the more perfectly formed annulose animals, this double chain becomes more or less completely united into a single chain. Mr. Newport has described the course of this concentration as exhibited in insects; and by Rathke it has been traced in crustaceans. During the early stages of the *Astacus fluviatilis*, or common cray-fish, there is a pair of separate ganglia to each ring. Of the fourteen pairs belonging to the head and thorax, the three pairs in advance of the mouth consolidate into one mass to form the brain, or cephalic ganglion. Meanwhile, out of the remainder, the first six pairs severally unite in the median line, while the rest remain more or less separate. Of these six double ganglia thus formed, the anterior four coalesce into one mass; the remaining two coalesce into another mass; and then these two masses coalesce into one. Here we see longitudinal and transverse integration going on simultaneously; and in the highest crustaceans they are both carried still further. The *Vertebrata* clearly exhibit transverse integration in the development of the generative system. The lowest of the mammalia—the *Monotremata*—in common with birds, to which they are in many respects allied, have oviducts which towards their lower extremities are dilated into cavities, severally performing in an imperfect way the function of a uterus. "In the *Marsupialia* there is a closer approximation of the two lateral sets of organs on the median line; for the oviducts converge towards one another and meet (without coalescing) on the median line; so that their uterine dilatations are in contact with each other, forming a true 'double uterus....' As we ascend the series of 'placental' mammals, we find the lateral coalescence becoming more and more complete.... In many of the *Rodentia* the uterus still remains completely divided into two lateral halves; whilst in others these coalesce at their lower portions, forming a rudiment of the true 'body' of the uterus in the human subject. This part increases at the expense of the lateral 'cornua' in the higher herbivora and carnivora; but even in the lower quadrumana the uterus is somewhat cleft at its summit."[10]

In the social organism integrative changes are not less clearly and abundantly exemplified. Uncivilized societies display them when wandering families, such as the bushmen show us, unite into tribes of considerable numbers. Among these we see a further progress of like nature everywhere manifested in the subjugation of weaker tribes by stronger ones; and in the subordination of their respective chiefs to the conquering chief. The partial combinations thus resulting, which among aboriginal races are being continually formed and continually broken up, become, among the superior races, both more complete and more permanent. If we trace the metamorphoses through which our own society, or any adjacent one, has passed, we see this unification from time to time repeated on a larger scale and with increasing stability. The aggregation of juniors and the children of juniors under elders and the children of elders; the consequent establishment of groups of vassals bound to their respective nobles; the subordination afterwards established of groups of inferior nobles to dukes or earls; and the still later establishment of the kingly power over dukes or earls; are so many instances of increasing consolidation. This process through which petty tenures are combined into feuds, feuds into provinces, provinces into kingdoms, and finally contiguous kingdoms into a single one, slowly completes itself by destroying the original lines of demarcation. And it may be further remarked of the European nations as a whole, that in the tendency to form alliances more or less lasting, in the restraining influences exercised by the several governments over each other, in the system that is gradually establishing itself of settling international disputes by congresses, as well as in the breaking down of commercial barriers and the increasing facilities of communication, we may trace the incipient stage of a European confederation—a still larger integration than any now established. But it is not only in these external unions of groups with groups, and of the compound groups with each other, that the general law is exemplified. It is exemplified also in unions that take place internally, as the groups become more highly organized. These, of which the most conspicuous are commercial in their origin and function, are well illustrated in our own society. We have integrations consequent on the simple growth of adjacent parts performing like functions: as, for instance, the junction of Manchester with its calico-weaving suburbs. We have other integrations that arise when, out of several places producing a particular commodity, one monopolizes more and more of the business, and leaves the rest to dwindle: as witness the growth of the Yorkshire cloth-districts at the expense of those in the west of England; or the absorption by Staffordshire of the pottery-manufacture, and the consequent decay of the establishments that once flourished at Worcester, Derby, and elsewhere. And we have those yet other integrations produced by the actual approximation of the similarly-occupied parts: whence result such facts as the concentration of publishers in Paternoster Row; of lawyers in the Temple and neighbourhood; of corn-merchants about Mark Lane; of civil engineers in Great George Street; of bankers in the centre of the city. Industrial combinations that consist, not in the approximation or fusion of parts, but in the establishment of common centres of connexion, are exhibited in the Bank clearing-house and the Railway clearing-house. While of yet another genus are those unions which bring into relation the more or less dispersed citizens who are occupied in like ways: as traders are brought by the Exchange and the Stock-Exchange; and as are professional men by institutes, like those of Civil Engineers, Architects, &c.

Here, as before, it is manifest that a law of Evolution which holds of organisms, must hold too of all objective results of their activity; and that hence Language, and Science, and Art, must not only in the course of their development display increasing heterogeneity and definiteness, but also increasing integration. We shall find this conclusion to be in harmony with the facts.

Among uncivilized races, the many-syllabled terms used for not uncommon objects, as well as the descriptive character of proper names, show us that the words used for the less familiar things are formed by compounding the words used for the more familiar things. This process of composition is sometimes found in its incipient stage—a stage in which the component words are temporarily united to signify some unnamed object, and do not (from lack of frequent use) permanently cohere. But in the majority of inferior languages, the process of "agglutination," as it is called, has gone far enough to produce considerable stability in the compound words: there is a manifest integration. How small is the degree of this integration, however, when compared with that reached in well-developed languages is shown both by the great length of the compound words used for things and acts of constant occurrence, and by the separableness of their elements. Certain North-American tongues very well illustrate this. In a Ricaree vocabulary extending to fifty names of common objects, which in English are nearly all expressed by single syllables, there is not one monosyllabic word; and in the nearly-allied vocabulary of the Pawnees, the names for these same common objects are monosyllabic in but two instances. Things so familiar to these hunting tribes as *dog* and *bow*, are, in the Pawnee language, *ashakish* and *teeragish*; the *hand* and the *eyes* are respectively *iksheeree* and *keereekoo*; for *day* the term is *shakoorooeeshairet*, and for *devil* it is *tsaheekshkakooraiwah*; while the numerals are composed of from two syllables up to five, and in Ricaree up to seven. That the great length of these familiar words implies a low degree of development, and that in the formation of higher languages out of lower there is a progressive integration, which reduces the polysyllables to dissyllables and monosyllables, is an inference fully confirmed by the history of our own language. Anglo-Saxon *steorra* has been in course of time consolidated into English *star*, *mona* into *moon*, and *nama* into *name*. The transition through the intermediate semi-Saxon is clearly traceable. *Sunu* became in semi-Saxon *sune*, and in English *son*: the final *e* of *sune* being an evanescent form of the original *u*. The change from the Anglo-Saxon plural, formed by the distinct syllable *as*, to our plural formed by the appended consonant *s*, shows us the same thing: *smithas* in becoming *smiths*, and *endas* in becoming *ends*, illustrate progressive coalescence. So too does the disappearance of the terminal *an* in the infinitive mood of verbs; as shown in the transition from the Anglo-Saxon *cuman* to the semi-Saxon *cumme*, and to the English *come*. Moreover the process has been slowly going on, even since what we distinguish as English was formed. In Elizabeth's time, verbs were still very frequently pluralized by the addition of *en*—we *tell* was we *tellen*; and in some rural districts this form of speech may even now be heard. In like manner the terminal *ed* of the past tense, has united with the word it modifies. *Burn-ed* has in pronunciation become *burnt*; and even in writing the terminal *t* has in some cases taken the place of the *ed*. Only where antique forms in general are adhered to, as in the church-service, is the distinctness of this inflection still maintained. Further, we see that the compound vowels have been in many cases fused into single vowels. That in *bread* the *e* and *a* were originally both sounded, is proved by the fact that they are still so sounded in parts where old habits linger. We, however, have contracted the pronunciation into *bred*; and we have made like changes in many other common words. Lastly, let it be noted that where the frequency of repetition is greatest, the process is carried furthest; as instance the contraction of *lord* (originally *laford*) into *lud* in the mouths of Barristers; and still better the coalescence of *God be with you* into *Good bye*. Besides exhibiting in this way the integrative process, Language equally exhibits it throughout all grammatical development. The lowest kinds of human speech, having merely nouns and verbs without inflections to them, manifestly permit no such close union of the elements of a proposition as results when the relations are either marked by inflections or by words specially used for purposes of connexion. Such speech is necessarily what we significantly call "incoherent." To a considerable extent, incoherence is seen in the Chinese language. "If, instead of saying *I go* to *London, figs come* from *Turkey, the sun shines* through *the air*, we said, *I go* end *London, figs come* origin *Turkey, the sun shines* passage *air*, we should discourse of the manner of the Chinese." From this "aptotic" form, there is clear evidence of a transition by coalescence to a form in which the connexions of words are expressed by the addition to them of certain inflectional words. "In Languages like the Chinese," remarks Dr Latham, "the separate words most in use to express relation may become adjuncts or annexes." To this he adds the fact that "the numerous inflexional languages fall into two classes. In one, the inflexions have no appearance of having been separate words. In the other, their origin as separate words is demonstrable." From which the inference drawn is, that the "aptotic" languages, by the more and more constant use of adjuncts, gave rise to the "agglutinate" languages, or those in which the original separateness of the inflexional parts can be traced; and that out of these, by further use, arose the "amalgamate" languages, or these in which the original separateness of the inflexional parts can no longer be traced. Strongly corroborative of this inference is the unquestionable fact, that by such a process there have grown out of the amalgamate languages, the "anaptotic" languages; of which our own is the most perfect example—languages in which, by further consolidation, inflexions have almost disappeared, while, to express the verbal relations, certain new kinds of words have been developed. When we see the Anglo-Saxon inflexions gradually lost by contraction during the development of English, and, though to a less degree, the Latin inflexions dwindling away during the development of French, we cannot deny that grammatical structure is modified by integration; and seeing how clearly the earlier stages of grammatical structure are explained by it, we can scarcely doubt that it has been going on from the first. And now mark that in proportion to the degree of the integration above described, is the extent to which integration of another order is shown. Aptotic languages are, as already pointed out, necessarily incoherent—the elements of a proposition cannot be tied into a definite and complete whole. But as fast as coalescence produces inflected words, it becomes possible to unite them into sentences of which the parts are so mutually dependent that no considerable change can be made without destroying the meaning. Yet a further stage in this process may be noted. After the development of those grammatical forms which make definite statements possible, we do not at first find them used to express anything beyond statements of a simple kind. A single subject with a single predicate, accompanied by but few qualifying terms, are usually all. If we compare, for instance, the Hebrew scriptures with writings of modern times, a marked difference of aggregation among the groups of words, is visible. In the number of subordinate propositions which accompany the principal one; in the various complements to subjects and predicates; and in the numerous qualifying clauses—all of them united into one complex whole—many sentences in modern composition exhibit a degree of integration not to be found in ancient ones.

The history of Science presents facts of the same meaning at every step. Indeed the integration of groups of like entities and like relations, may be said to constitute the most conspicuous part of scientific progress. A glance at the classificatory sciences, shows us not only that the confused aggregations which the vulgar make of natural objects, are differentiated into groups that are respectively more homogeneous, but also that these groups are gradually rendered complete and compact. While, instead of considering all marine creatures as fish, shell-fish, and jelly-fish, Zoology establishes divisions and sub-divisions under the heads *Vertebrata, Annulosa, Mollusca,* &c.— while in place of the wide and vague assemblage popularly described as "creeping things," it makes the specific classes *Annelida, Myriopoda, Insecta, Arachnida;* it at the same time gives to these an increasing consolidation. The several orders and genera of which each consists, are arranged according to their affinities and bound together under common definitions; at the same time that, by extended observation and rigorous criticism, the previously unknown and undetermined forms are integrated with their respective congeners. Nor is the same process less clearly manifested in those sciences which have for their subject-matter, not classified objects, but classified relations. Under one of its chief aspects, the advance of Science is the advance of generalization; and generalization is the uniting into groups all like co-existencies and sequences among phenomena. Not only, however, does the colligation of a number of concrete relations into a generalization of the lowest order, exemplify the principle enunciated; but it is again and again exemplified in the colligation of these lowest generalizations into higher ones, and these into still higher ones. Year by year are established certain connexions among orders of phenomena that seem wholly unallied;

and these connexions, multiplying and strengthening, gradually bring the seemingly unallied orders under a common bond. When, for example, Humboldt quotes the saying of the Swiss—"it is going to rain because we hear the murmur of the torrents nearer,"—when he remarks the relation between this and an observation of his own, that the cataracts of the Orinoco are heard at a greater distance by night than by day—when he notes the essential parallelism existing between these facts and the fact that the unusual visibility of remote objects is also an indication of coming rain—and when he points out that the common cause of these variations is the smaller hindrance offered to the passage of both light and sound, by media which are comparatively homogeneous, either in temperature or hygrometric state; he helps in bringing under one generalization the phenomena of light and those of sound. Experiment having shown that these conform to like laws of reflection and refraction, the conclusion that they are both produced by undulations gains probability: there is an incipient integration of two great orders of phenomena, between which no connexion was suspected in times past. A still more decided integration has been of late taking place between the once independent sub-sciences of Electricity, Magnetism, and Light. And indeed it must be obvious to those who are familiar with the present state of Science, that there will eventually take place a far wider integration, by which all orders of phenomena will be combined as differently conditioned forms of one ultimate fact.

Nor do the industrial and æsthetic Arts fail to supply us with equally conclusive evidence. The progress from rude, small, and simple tools, to perfect, complex, and large machines, illustrates not only a progress in heterogeneity and in definiteness, but also in integration. Among what are classed as the mechanical powers, the advance from the lever to the wheel-and-axle is an advance from a simple agent to an agent made up of several simple ones combined together. On comparing the wheel-and-axle, or any of the machines used in early times with those used now, we find an essential difference to be, that in each of our machines several of the primitive machines are united into one. A modern apparatus for spinning or weaving, for making stockings or lace, contains not simply a lever, an inclined plane, a screw, a wheel-and-axle, united together; but several of each integrated into one complex whole. Again, in early ages, when horse-power and man-power were alone employed, the motive agent was not bound up with the tool moved; but the two have now become in many cases fused together: the fire-box and boiler of a locomotive are combined with the machinery which the steam works. Nor is this the most extreme case. A still more extensive integration is exhibited in every large factory. Here we find a large number of complicated machines, all connected by driving shafts with the same steam-engine—all united with it into one vast apparatus. Contrast the mural decorations of the Egyptians and Assyrians with modern historical paintings, and there becomes manifest a great advance in unity of composition—in the subordination of the parts to the whole. One of these ancient frescoes is in truth made up of a number of pictures that have little mutual dependence. The several figures of which each group consists, show very imperfectly by their attitudes, and not at all by their expressions, the relations in which they stand to each other; the respective groups might be separated with but little loss of meaning; and the centre of chief interest, which should link all parts together, is often inconspicuous. The same trait may be noted in the tapestries of medieval days. Representing perhaps a hunting scene, one of these exhibits men, horses, dogs, beasts, birds, trees, and flowers, miscellaneously dispersed: the living objects being variously occupied, and mostly with no apparent consciousness of each other's proximity. But in the paintings since produced, faulty as many of them are in this respect, there is always a more or less manifest co-ordination of parts—an arrangement of attitudes, expressions, lights, and colours, such as to combine the picture into an organic whole; and the success with which unity of effect is educed from variety of components, is a chief test of merit. In music, progressive integration is displayed in still more numerous ways. The simple cadence embracing but a few notes, which in the chants of savages is monotonously repeated, becomes among civilized races, a long series of different musical phrases combined into one whole; and so complete is the integration, that the melody cannot be broken off in the middle, nor shorn of its final note, without giving us a painful sense of incompleteness. When to the air, a bass, a tenor, and an alto are added; and when to the harmony of different voice-parts there is added an accompaniment; we see exemplified integrations of another order, which grow gradually more elaborate. And the process is carried a stage higher when these complex solos, concerted pieces, choruses, and orchestral effects, are combined into the vast ensemble of a musical drama; of which, be it remembered, the artistic perfection largely consists in the subordination of the particular effects to the total effect. Once more the Arts of literary delineation, narrative and dramatic, furnish us with parallel illustrations. The tales of primitive times, like those with which the story-tellers of the East still daily amuse their listeners, are made up of successive occurrences that are not only in themselves unnatural, but have no natural connexion: they are but so many separate adventures put together without necessary sequence. But in a good modern work of imagination, the events are the proper products of the characters working under given conditions; and cannot at will be changed in their order or kind, without injuring or destroying the general effect. And further, the characters themselves, which in early fictions play their respective parts without showing us how their minds are modified by each other or by the events, are now presented to us as held together by complex moral relations, and as acting and re-acting upon each other's natures.

Evolution, then, is in all cases a change from a more diffused or incoherent form, to a more consolidated or coherent form. This proves to be a characteristic displayed equally in those earliest changes which the Universe as a whole is supposed to have undergone, and in those latest changes which we trace in society and the products of social life. Nor is it only that in the development of a planet, of an organism, of a society, of a science, of an art, the process of integration is seen in a more complete aggregation of each whole and of its constituent parts; but it is also shown in an increasing mutual dependence of the parts. Dimly foreshadowed as this mutual dependence is among inorganic phenomena, both celestial and terrestrial, it becomes distinct among organic phenomena. From the lowest living forms upwards, the degree of development is marked by the degree in which the several parts constitute a mutually-dependent whole. The advance from those creatures which live on in each part when cut in pieces, up to those creatures which cannot lose any considerable part without death, nor any inconsiderable part without great constitutional disturbance, is clearly an advance to creatures which are not only more integrated in respect of their solidification, but are also more integrated as consisting of organs that live for and by each other. The like contrast between undeveloped and developed societies, need not be shown in detail: the ever-increasing co-ordination of parts, is conspicuous to all. And it must suffice just to indicate that the same thing holds true of social products: as, for instance, of Science; which has become highly integrated not only in the sense that each division is made up of mutually-dependent propositions, but also in the sense that the several divisions are mutually-dependent— cannot carry on their respective investigations without aid from each other.

It seems proper to remark that the generalization here variously illustrated, is akin to one enunciated by Schelling, that Life is the tendency to individuation. Struck by the fact that an aggregative process is traceable throughout nature, from the growth of a crystal up to the development of a man; and by the fact that the wholes resulting from this process, completer in organic than in inorganic bodies, are completest where the vital manifestations are the highest; Schelling concluded that this characteristic was the essential one. According to him, the formation of individual bodies is not incident to Life, but is that in which Life fundamentally consists. This position is, for several reasons, untenable. In the first place, it requires the conception of Life to be extended so as to embrace inorganic phenomena; since in crystallization, and even in the formation of amorphous masses of matter, this tendency to individuation is displayed. Schelling, fully perceiving this, did indeed accept the implication; and held that inorganic bodies had life lower only in degree than that of organic bodies—their degree of life being measured by their degree of individuation. This bold assumption, which Schelling evidently made to save his definition, is inadmissible. Rational philosophy cannot ignore those broad distinctions which the general sense of mankind has established. If it transcends them, it must at the same time show what is their origin; how far only they are valid; and why they disappear from a higher point of view. Note next that the more complete individuality which Schelling pointed out as characterizing bodies having the greatest amount of life, is only *one* of their structural traits. The greater degree of heterogeneity which they exhibit, is, as we have seen, a much more conspicuous peculiarity; and though it might possibly be contended that greater heterogeneity is remotely implied by greater individuality, it must be admitted that in defining Life as the tendency to individuation, no hint is given that the bodies which live most are the most heterogeneous bodies. Moreover it is to be remarked that this definition of Schelling, refers much more to the structures of living bodies than to the processes which constitute Life. Not Life, but the invariable accompaniment of Life, is that which his formula alone expresses. The formation of a completer organic whole, a more fully individuated body, is truly a necessary concomitant of a higher life; and the development of a higher life must therefore be accompanied by a tendency to greater individuation. But to represent this tendency as Life itself, is to mistake an incidental result for an original cause. Life, properly so called, consists of multiform changes united together in various ways; and is not expressed either by an anatomical description of the organism which manifests it, or by a history of the modifications through which such organism has reached its present structure. Yet it is only in such description and such history that the tendency to individuation is seen. Lastly, this definition which Schelling gave of Life is untenable, not only because it refers rather to the organism than to the actions going on in it; but also because it wholly ignores that connexion between the organism and the external world, on which Life depends. All organic processes, physical and psychial, having for their object the maintenance of certain relations with environing agencies and objects; it is impossible that there should be a true definition of Life, in which the environment is not named. Nevertheless, Schelling's conception was not a baseless one. Though not a truth, it was yet the adumbration of a truth. In defining Life as the tendency to individuation, he had in view that formation of a more compact, complete, and mutually-dependent whole, which, as we have seen, is one characteristic of Evolution in general. His error was, firstly, in regarding it as a characteristic of Life, instead of a characteristic of living bodies, displayed, though in a less degree, by other bodies; and, secondly, in regarding it as the sole characteristic of such bodies. It remains only to add, that for expressing this aspect of the process of Evolution, the word integration is for several reasons preferable to the word individuation. Integration is the true antithesis of differentiation; it has not that tacit reference to living bodies which the word individuation cannot be wholly freed from; it expresses the aggregative tendency not only as displayed in the formation of more complete wholes, but also as displayed in the consolidation of the several parts of which such wholes are made up; and it has not the remotest teleological implication. In short, it simply formulates in the most abstract manner, a wide induction untainted by any hypothesis.

§ 57. Thus we find that to complete the definition arrived at in the last chapter, much has to be added. What was there alleged is true; but it is not the whole truth. Evolution is unquestionably a change from a homogeneous state to a heterogeneous state; but, as we have seen, there are some advances in heterogeneity which cannot be included in the idea of Evolution. This undue width of the definition, implies the omission of some further peculiarity by which Evolution is distinguished; and this peculiarity we find to be that the more highly developed things become, the more definite they become. Advance from the indefinite to the definite, is as constantly and variously displayed as advance from the homogeneous to the heterogeneous. And we are thus obliged to regard it as an essential characteristic of Evolution. Further analysis, however, shows us that this increase of definiteness is not an independent process; but is rather the necessary concomitant of another process. A very little consideration of the facts proves that a change from the indefinite to the definite, can arise only through a completer consolidation of the respective parts, and of the whole which they constitute. And so we find that while Evolution is a transformation of the homogeneous into the heterogeneous, and of the indefinite into the definite, it is also a transformation of the incoherent into the coherent. Along with the differentiation shown in increasing contrasts of parts with each other, there goes on an integration, by which the parts are rendered distinct units, as well as closely united components of one whole. These clauses here added to the definition, are essential ones; not only as being needful to distinguish Evolution from that which is not Evolution, but likewise as being needful to express all which the idea of Evolution includes. Progressive integration with the growing definiteness necessarily resulting from it, is of co-ordinate importance with the progressive differentiation before dwelt upon—nay, from one point of view, may be held of greater importance. For organization, in which what we call Evolution is most clearly and variously displayed, consists even more in the union of many parts into one whole, than in the formation of many parts. The Evolution which we see throughout inorganic nature, is lower than that which organic nature exhibits to us, for the especial reason that the mutual dependence of parts is extremely indefinite, even when traceable at all. In an amorphous mass of matter, you may act mechanically or chemically upon one part without appreciably affecting the other parts. Though their electrical or thermal states may be for the moment altered, their original states are soon resumed. Even in the highest inorganic aggregation—a crystal—the apex may be broken off and leave the rest intact: the only clear evidence of mutual dependence of parts, being, the ability of the crystal to regenerate its apex if replaced in the solution from which it was formed. But the constituent parts of organic bodies can severally maintain their existing states, only while remaining in connexion. Even in the lowest living forms, mutilation cannot be carried beyond a certain point without decomposition ensuing. As we advance through the higher up to the highest forms, we see a gradual narrowing of the limits within which the mutilation does not cause destruction: a progressive increase of mutual dependence or integration which is, at the same time, the condition to greater functional perfection. In societies this truth is equally manifest. That the component units slowly segregate into groups of different ranks and occupations, is a fact scarcely more conspicuous than is the fact that these groups are necessary to each other's existence. And we cannot contemplate the still-progressing division of labour, without seeing that the interdependence becomes ever greater as the evolution becomes higher. It remains only to point out definitely, what has been already implied, that these several forms of change which have been successively described as making up the process of Evolution, are not in reality separate forms of change, but different aspects of the same change. Intrinsically the transformation is one and indivisible. The establishment of differences that become gradually more decided, is evidently but the beginning of an action which cannot be pushed to its extreme without producing definite divisions between the parts, and reducing each part to a separate mass. But with our limited faculties, it is not possible to take in the entire process at one view; nor have we any single terms by which the process can be described. Hence we are obliged to contemplate each of its aspects separately, and to find a separate expression for its characteristic.

Having done this, we are now in a position to frame a true idea of Evolution. Combining these partial definitions we get a complete definition, which may be most conveniently expressed thus—*Evolution is a change from an indefinite, incoherent homogeneity, to a definite, coherent heterogeneity; through continuous differentiations and integrations.*

It may perhaps be remarked that the last of these clauses is superfluous; since the differentiation and integration are implied in the first clause. This is true: the transition which the first clause specifies, is impossible save through the process specified in the second. Nevertheless, a mere statement of the two extreme stages with which Evolution begins and ends, omitting all reference to changes connecting them, leaves the mind with but an incomplete idea. The idea becomes much more concrete when these changes are described. Hence, though not logically necessary, the second clause of the definition is practically desirable.

Before closing the chapter, a few words must be added respecting certain other modes of describing Evolution. Organic bodies, from the changes of which the idea of Evolution has arisen, and to the changes of which alone it is usually applied, are often said to progress from simplicity to complexity. The transformation of the simple into the complex, and of the homogeneous into the heterogeneous, are used as equivalent phrases; or, if any difference is recognized between them, it is to the advantage of the first, which is held to be the more specific. After what has been said, however, it must be obvious that Evolution cannot be thus adequately formulated. No hint is given of that increased definiteness which we have found to be a concomitant of development. Nor is there anything implying the greater mutual dependence of parts. Nevertheless, the brevity of the expression gives it a value for ordinary purposes; and I shall probably hereafter frequently use it, both in those cases where more precise language is not demanded, and in those cases where it indicates the particular aspect of Evolution referred to. Another description frequently given of Evolution, is, that it is a change from the general to the special. The more or less spherical germ from which every organism, animal and vegetal, proceeds, is comparatively general: alike in the sense that in appearance and chemical nature it is very similar to all other germs; and also in the sense that its form is less markedly distinguished from the average forms of objects at large, than is that of the mature organism—a contrast which equally holds of internal structure. But this progress from the more general to the more special, is rather a derivative than an original characteristic. An increase of speciality being really an increase in the number of attributes—an addition of traits not possessed by bodies that are in other respects similar—is a necessary result of multiplying differentiations. In other words, general and special are subjective or ideal distinctions involved in our conceptions of classes, rather than objective or real distinctions presented in the bodies classified. Nevertheless, this abstract formula is not without its use. It expresses a fact of much significance; and one which we shall have constantly to refer to when dealing with the relations between organic bodies and their surrounding conditions.

The law of Evolution however, be it expressed in full as above, or in these shorter but less specific phrases, is essentially that which has been exhibited in detail throughout the foregoing pages. So far as we can ascertain, this law is universal. It is illustrated with endless repetition, and in countless ways, wherever the facts are abundant; and where the facts do not suffice for induction, deduction goes far to supply its place. Among all orders of phenomena that lie within the sphere of observation, we see ever going on the process of change above defined; and many significant indications warrant us in believing, that the same process of change went on throughout that remote past which lies beyond the sphere of observation. If we must form any conclusion respecting the general course of things, past, present, and future, the one which the evidence as far as it goes justifies, and the only one for which there is any justification, is, that the change from an indeterminate uniformity to a determinate multiformity which we everywhere see going on, has been going on from the first, and will continue to go on.

10. Carpenter's Prin. of Comp. Phys., p. 617.

CHAPTER IV
THE CAUSES OF EVOLUTION

§ 58. Is this law ultimate or derivative? Must we rest satisfied with the conclusion that throughout all classes of concrete phenomena such is the mode of evolution? Or is it possible for us to ascertain *why* such is the mode of evolution? May we seek for some all-pervading principle which underlies this all-pervading process? Can we by a further step reduce our empirical generalization to a rational generalization?

Manifestly this community of result implies community of causation. It may be that of such causation no account can be given, further than that the Unknowable is manifested to us after this manner. Or, it may be, that the mode of manifestation is reducible to simpler ones, from which these many complex consequences follow. Analogy suggests the latter inference. At present, the conclusion that every kind of Evolution is from a state of indefinite incoherent homogeneity to a state of definite coherent heterogeneity, stands in the same position as did the once ultimate conclusion that every kind of organized body undergoes, when dead, a more or less rapid decay. And as, for the various kinds of decomposition through which animal and vegetal products pass, we have now discovered a rationale in the chemical affinities of their constituent elements; so, possibly, this universal transformation of the simple into the complex may be affiliated upon certain simple primordial principles.

Such cause or causes of Evolution, may be sought for without in the least assuming that the ultimate mystery can be fathomed. Fully conscious that an absolute solution is for ever beyond us, we may still look for a relative solution—may try to reduce the problem to its lowest terms. Just as it was possible to interpret Kepler's laws as necessary consequences of the law of gravitation, and then to admit that gravitation transcends analysis; so it may be possible to interpret the law of Evolution as the necessary consequence of some deeper law, beyond which we may nevertheless be unable to go.

§ 59. The probability of common causation, and the possibility of formulating it, being granted, it will be well before going further, to consider what must be the general characteristics of such causation, and in what direction we ought to look for it. We can with certainty predict that be it simple or compound, the cause has a high degree of generality; seeing that it is common to such infinitely varied phenomena: in proportion to the universality of its application must be the abstractness of its character. Whatever be the agency and the conditions under which it acts, we need not expect to see in them an obvious explanation of this or that species of Evolution, because they equally underlie species of Evolution of quite a different order. Determining Evolution of every kind—astronomic, geologic, organic, ethnologic, social, economic, artistic, &c.—they must be concerned with something common to all these; and to see what these possess in common, will therefore be the best method of guiding ourselves towards the desired solution.

The only obvious respect in which all kinds of Evolution are alike, is, that they are modes of *change*. Every phenomenon to which we apply the term, presents us with a succession of states; and when such succession ceases, we no longer predicate Evolution. Equally in those past forms of it which are more or less hypothetical, and in those forms of it which we see going on around, this is the common characteristic. Note next, that the kind of change which constitutes Evolution, is broadly distinguished from change of an equally general kind, in this, that it is change of internal relations instead of change of external relations. All things in motion through space are the subjects of change; but while in this which we call mechanical motion, the relative position as measured from surrounding objects is continually altered, there is not implied any alteration in the positions of the parts of the moving body in respect to each other. Conversely, a body exhibiting what we call Evolution, while it either may or may not display new relations of position to the things around it, *must* display new relations of position among the parts of which it is made up. Thus we narrow the field of inquiry by recognizing the change in which Evolution consists, as *a change in the arrangement of parts*: of course using the word parts in its most extended sense, as signifying both ultimate units and masses of such units. Further, we have to remember that this change in the arrangement of parts which constitutes Evolution, is a certain order of such change. As we saw in the last chapter, there is a change in the arrangement of parts which is not Evolution but Dissolution—a destructive change as opposed to a constructive change—a change by which the definite is gradually rendered indefinite, the coherent slowly becomes incoherent, and the heterogeneous eventually lapses into comparative homogeneity. Thus then we reduce that which we have to investigate to its most abstract shape. Our task is to find the cause or causes of a certain order of change that takes place in the arrangement of parts.

§ 60. Evidently the problem, as thus expressed, brings us face to face with the ultimate elements of phenomena in general. It is impossible to account for a certain change in the arrangement of the parts of any mass, without involving—first, the *matter* which makes up the parts thus re-arranged; next, the *motion* exhibited during the re-arrangement; and then, the *force* producing this motion. The problem is a dynamical one; and there can be no truly scientific solution of it, save one given in terms of Matter, Motion, and Force—terms in which all other dynamical problems are expressed and solved.

The proposal thus to study the question from a purely physical point of view, will most likely, notwithstanding what has been said in the first part of this work, raise in some minds either alarm or prejudice. Having, throughout life, constantly heard the charge of materialism made against those who ascribed the more involved phenomena to agencies like those seen in the simplest phenomena, most persons have acquired a repugnance to such methods of interpretation; and when it is proposed to apply them universally, even though it is premised that the solution they give can be but relative, more or less of the habitual feeling will probably arise. Such an attitude of mind, however, is significant, not so much of a reverence for the Unknown Cause, as of an irreverence for those omnipresent forms in which the Unknown Cause is manifested to us. Men who have not risen above that vulgar conception which unites with Matter the contemptuous epithets "gross" and "brute," may naturally enough feel dismay at the proposal to reduce the phenomena of Life, of Mind, and of Society, to a level with those which they think so degraded. But whoever remembers that the forms of existence which the uncultivated speak of with so much scorn, are not only shown by the man of science to be the more marvellous in their attributes the more they are investigated, but are also proved to be in their ultimate nature absolutely incomprehensible—as absolutely incomprehensible as sensation, or the conscious something which perceives it—whoever fully realizes this truth, I say, will see that the course proposed does not imply a degradation of the so-called higher, but an elevation of the so-called lower. Perceiving, as he will, that the Materialist and Spiritualist controversy is a mere war of words, in which the disputants are equally absurd—each thinking he understands that which it is impossible for any man to understand—he will perceive how utterly groundless is the fear referred to. Being fully convinced that whatever nomenclature is used, the ultimate mystery must remain the same, he will be as ready to formulate all phenomena in terms of Matter, Motion, and Force, as in any other terms; and will rather indeed anticipate, that only in a doctrine which recognizes the Unknown Cause as co-extensive with all orders of phenomena, can there be a consistent Religion, or a consistent Science.

On the other hand, the conclusion that Evolution, considered under its most abstract form, is a certain change in the arrangement of parts; and that the causes of this change can be expressed only in terms of Matter, Motion, and Force; may in critical minds raise the question—What are Matter, Motion, and Force? Referring back in thought to the reasonings contained in the chapter on "Ultimate Scientific Ideas;" and remembering how it was there shown that absolute knowledge of Matter, Motion, and Force, is impossible; some readers will perhaps conclude that any such interpretation as the one above proposed, must be visionary. It may be asked—How can a comprehensible account of Evolution be given in terms that are themselves incomprehensible?

Before proceeding, this question must be met. There can be no sound philosophy without clearly defined terms; and as, on the meanings of the terms to be here used, doubts have probably been cast by the reasonings contained in the chapter referred to, such doubts must be removed. If, as was shown, our ideas of things do not correspond with things in themselves, it becomes necessary to inquire in what way they are to be accepted. If they are not absolutely true, then what is the exact meaning of the assertion that they are relatively true? To this question let us now address ourselves.

CHAPTER V
SPACE, TIME, MATTER, MOTION, AND FORCE

§ 61. That sceptical state of mind which the criticisms of Philosophy usually produce, is, in great measure, caused by the misinterpretation of words. A sense of universal illusion ordinarily follows the reading of metaphysics; and is strong in proportion as the argument has appeared conclusive. This sense of universal illusion would probably never have arisen, had the terms used been always rightly construed. Unfortunately, these terms have by association acquired meanings that are quite different from those given to them in philosophical discussions; and the ordinary meanings being unavoidably suggested, there results more or less of that dreamlike idealism which is so incongruous with our instinctive convictions. The word *phenomenon* and its equivalent word *appearance*, are in great part to blame for this. In ordinary speech, these are uniformly employed in reference to visual perceptions. Habit, almost, if not quite, disables us from thinking of *appearance* except as something seen; and though *phenomenon* has a more generalized meaning, yet we cannot rid it of associations with *appearance*, which is its verbal equivalent. When, therefore, Philosophy proves that our knowledge of the external world can be but phenomenal—when it concludes that the things of which we are conscious are appearances; it inevitably arouses in us the notion of an illusiveness like that to which our visual perceptions are so liable in comparison with our tactual perceptions. Good pictures show us that the aspects of things may be very nearly simulated by colours on canvass. The looking-glass still more distinctly proves how deceptive is sight when unverified by touch. And the frequent cases in which we misinterpret the impressions made on our eyes, and think we see something which we do not see, further shake our faith in vision. So that the implication of uncertainty has infected the very word *appearance*. Hence, Philosophy, by giving it an extended meaning, leads us to think of all our senses as deceiving us in the same way that the eyes do; and so makes us feel ourselves floating in a world of phantasms. Had *phenomenon* and *appearance* no such misleading associations, little, if any, of this mental confusion would result. Or did we in place of them use the term *effect*, which is equally applicable to all impressions produced on consciousness through any of the senses, and which carries with it in thought the necessary correlative *cause*, with which it is equally real, we should be in little danger of falling into the insanities of idealism.

Such danger as there might still remain, would disappear on making a further verbal correction. At present, the confusion resulting from the above misinterpretation, is made greater by an antithetical misinterpretation. We increase the seeming unreality of that phenomenal existence which we can alone know, by contrasting it with a noumenal existence which we imagine would, if we could know it, be more truly real to us. But we delude ourselves with a verbal fiction. What is the meaning of the word *real*? This is the question which underlies every metaphysical inquiry; and the neglect of it is the remaining cause of the chronic antagonisms of metaphysicians. In the interpretation put on the word *real*, the discussions of philosophy retain one element of the vulgar conception of things, while they reject all its other elements; and create confusion by the inconsistency. The peasant, on contemplating an object, does not regard that which he contemplates as something in himself, but believes the thing of which he is conscious to be the external object—imagines that his consciousness extends to the very place where the object lies: to him the appearance and the reality are one and the same thing. The metaphysician, however, is convinced that consciousness cannot embrace the reality, but only the appearance of it; and so he transfers the appearance into consciousness and leaves the reality outside. This reality left outside of consciousness, he continues to think of much in the same way as the ignorant man thinks of the appearance. Though the reality is asserted to be out of consciousness, yet the *realness* ascribed to it is constantly spoken of as though it were a knowledge possessed apart from consciousness. It seems to be forgotten that the conception of reality can be nothing more than some mode of consciousness; and that the question to be considered is—What is the relation between this mode and other modes?

By reality we mean *persistence* in consciousness: a persistence that is either unconditional, as our consciousness of space, or that is conditional, as our consciousness of a body while grasping it. The real, as we conceive it, is distinguished solely by the test of persistence; for by this test we separate it from what we call the unreal. Between a person standing before us, and the idea of such a person, we discriminate by our ability to expel the idea from consciousness, and our inability, while looking at him, to expel the person from consciousness. And when in doubt as to the validity or illusiveness of some impression made upon us in the dusk, we settle the matter by observing whether the impression persists on closer observation; and we predicate reality if the persistence is complete. How truly persistence is what we mean by reality, is shown in the fact that when, after criticism has proved that the real as we are conscious of it is not the objectively real, the indefinite notion which we form of the objectively real, is of something which persists absolutely, under all changes of mode, form, or appearance. And the fact that we cannot form even an indefinite notion of the absolutely real, except as the absolutely persistent, clearly implies that persistence is our ultimate test of the real as present to consciousness.

Reality then, as we think it, being nothing more than persistence in consciousness, the result must be the same to us whether that which we perceive be the Unconditioned itself, or an effect invariably wrought on us by the Unconditioned. If some mode of the Unconditioned uniformly produces some mode of consciousness—if the mode of consciousness so produced, is as persistent as would be such mode of the Unconditioned were it immediately known; it follows that the reality will be to our consciousness as complete in the one case as in the other. Were the Unconditioned itself present in thought, it could but be persistent; and if instead of it, there is present its persistent effect, the resulting consciousness of reality must be exactly the same.

Hence there may be drawn these conclusions:—First, that we have an indefinite consciousness of an absolute reality transcending relations, which is produced by the absolute persistence in us of something which survives all changes of relation. Second, that we have a definite consciousness of relative reality, which unceasingly persists in us under one or other of its forms, and under each form so long as the conditions of presentation are fulfilled; and that the relative reality, being thus continuously persistent in us, is as real to us as would be the absolute reality could it be immediately known. Third, that thought being possible only under relation, the relative reality can be conceived as such only in connexion with an absolute reality; and the connexion between the two being absolutely persistent in our consciousness, is real in the same sense as the terms it unites are real.

Thus then we may resume, with entire confidence, those realistic conceptions which philosophy at first sight seems to dissipate. Though reality under the forms of our consciousness, is but a conditioned effect of the absolute reality, yet this conditioned effect standing in indissoluble relation with its unconditioned cause, and being equally persistent with it so long as the conditions persist, is, to the consciousness supplying those conditions, equally real. The persistent impressions being the persistent results of a persistent cause, are for practical purposes the same to us as the cause itself; and may be habitually dealt with as its equivalents. Somewhat in the same way that our visual perceptions, though merely symbols found to be the equivalents of tactual perceptions, are yet so identified with those tactual perceptions that we actually appear to see the solidity and hardness which we do but infer, and thus conceive as objects what are only the signs of objects; so, on a higher stage, do we deal with these relative realities as though they were absolutes instead of effects of the absolute. And we may legitimately continue so to deal with them as long as the conclusions to which they help us are understood as relative realities and not absolute ones.

This general conclusion it now remains to interpret specifically, in its application to each of our ultimate scientific ideas.

§ 62.[11] We think in relations. This is truly the form of all thought; and if there are any other forms, they must be derived from this. We have seen (Chap. iii. Part I.) that the several ultimate modes of being cannot be known or conceived as they exist in themselves; that is, out of *relation* to our consciousness. We have seen, by analyzing the product of thought, (§ 23,) that it always consists of *relations*; and cannot include anything beyond the most general of these. On analyzing the process of thought, we found that cognition of the Absolute was impossible, because it presented neither *relation*, nor its elements—difference and likeness. Further, we found that not only Intelligence but Life itself, consists in the establishment of internal *relations* in correspondence with external relations. And lastly, it was shown that though by the relativity of our thought we are eternally debarred from knowing or conceiving Absolute Being; yet that this very *relativity* of our thought, necessitates that vague consciousness of Absolute Being which no mental effort can suppress. That *relation* is the universal form of thought, is thus a truth which all kinds of demonstration unite in proving.

By the transcendentalists, certain other phenomena of consciousness are regarded as forms of thought. Presuming that relation would be admitted by them to be a universal mental form, they would class with it two others as also universal. Were their hypothesis otherwise tenable however, it must still be rejected if such alleged further forms are interpretable as generated by the primary form. If we think in relations, and if relations have certain universal forms, it is manifest that such universal forms of relations will become universal forms of our consciousness. And if these further universal forms are thus explicable, it is superfluous, and therefore unphilosophical, to assign them an independent origin. Now relations are of two orders— relations of sequence, and relations of co-existence; of which the one is original and the other derivative. The relation of sequence is given in every change of consciousness. The relation of co-existence, which cannot be originally given in a consciousness of which the states are serial, becomes distinguished only when it is found that certain relations of sequence have their terms presented in consciousness in either order with equal facility; while the others are presented only in one order. Relations of which the terms are not reversible, become recognized as sequences proper; while relations of which the terms occur indifferently in both directions, become recognized as co-existences. Endless experiences, which from moment to moment present both orders of these relations, render the distinction between them perfectly definite; and at the same time generate an abstract conception of each. The abstract of all sequences is Time. The abstract of all co-existences is Space. From the fact that in thought, Time is inseparable from sequence, and Space from co-existence, we do not here infer that Time and Space are original conditions of consciousness under which sequences and co-existences are known; but we infer that our conceptions of Time and Space are generated, as other abstracts are generated from other concretes: the only difference being, that the organization of experiences has, in these cases, been going on throughout the entire evolution of intelligence.

This synthesis is confirmed by analysis. Our consciousness of Space is a consciousness of co-existent positions. Any limited portion of space can be conceived only by representing its limits as co-existing in certain relative positions; and each of its imagined boundaries, be it line or plane, can be thought of in no other way than as made up of co-existent positions in close proximity. And since a position is not an entity—since the congeries of positions which constitute any conceived portion of space, and mark its bounds, are not sensible existences; it follows that the co-existent positions which make up our consciousness of Space, are not co-existences in the full sense of the word (which implies realities as their terms), but are the blank forms of co-existences, left behind when the realities are absent; that is, are the abstracts of co-existences. The experiences out of which, during the evolution of intelligence, this abstract of all co-existences has been generated, are experiences of individual positions as ascertained by touch; and each of such experiences involves the resistance of an object touched, and the muscular tension which measures this resistance. By countless unlike muscular adjustments, involving unlike muscular tensions, different resisting positions are disclosed; and these, as they can be experienced in one order as readily as another, we regard as co-existing. But since, under other circumstances, the same muscular adjustments do not produce contact with resisting positions, there result the same states of consciousness, minus the resistances—blank forms of co-existence from which the co-existent objects before experienced are absent. And from a building up of these, too elaborate to be here detailed, results that abstract of all relations of co-existence which we call Space. It remains only to point out, as a thing which we must not forget, that the experiences from which the consciousness of Space arises, are experiences of *force*. A certain correlation of the muscular forces we ourselves exercise, is the index of each position as originally disclosed to us; and the resistance which makes us aware of something existing in that position, is an equivalent of the pressure we consciously exert. Thus, experiences of forces variously correlated, are those from which our consciousness of Space is abstracted.

That which we know as Space being thus shown, alike by its genesis and definition, to be purely relative, what are we to say of that which causes it? Is there an absolute Space which relative Space in some sort represents? Is Space in itself a form or condition of absolute existence, producing in our minds a corresponding form or condition of relative existence? These are unanswerable questions. Our conception of Space is produced by some mode of the Unknowable; and the complete unchangeableness of our conception of it simply implies a complete uniformity in the effects wrought by this mode of the Unknowable upon us. But therefore to call it a necessary mode of the Unknowable, is illegitimate. All we can assert is, that Space is a relative reality; that our consciousness of this unchanging relative reality implies an absolute reality equally unchanging in so far as we are concerned; and that the relative reality may be unhesitatingly accepted in thought as a valid basis for our reasonings; which, when rightly carried on, will bring us to truths that have a like relative reality—the only truths which concern us or can possibly be known to us.

Concerning Time, relative and absolute, a parallel argument leads to parallel conclusions. These are too obvious to need specifying in detail.

§ 63. Our conception of Matter, reduced to its simplest shape, is that of co-existent positions that offer resistance; as contrasted with our conception of Space, in which the co-existent positions offer no resistance. We think of Body as bounded by surfaces that resist; and as made up throughout of parts that resist. Mentally abstract the co-existent resistances, and the consciousness of Body disappears; leaving behind it the consciousness of Space. And since the group of co-existing resistent positions constituting a portion of matter, is uniformly capable of giving us impressions of resistance in combination with various muscular adjustments, according as we touch its near, its remote, its right, or its left side; it results that as different muscular adjustments habitually indicate different co-existences, we are obliged to conceive every portion of matter as containing more than one resistent position—that is, as occupying Space. Hence the necessity we are under of representing to ourselves the ultimate elements of Matter as being at once extended and resistent: this being the universal form of our sensible experiences of Matter, becomes the form which our conception of it cannot transcend, however minute the fragments which imaginary subdivisions produce. Of these two inseparable elements, the resistance is primary, and the extension secondary. Occupied extension, or Body, being distinguished in consciousness from unoccupied extension, or Space, by its resistance, this attribute must clearly have precedence in the genesis of the idea. Such a conclusion is, indeed, an obvious corollary from that at which we arrived in the foregoing section. If, as was there contended, our consciousness of Space is a product of accumulated experiences, partly our own but chiefly ancestral—if, as was pointed out, the experiences from which our consciousness of Space is abstracted, can be received only through impressions of resistance made upon the organism; the necessary inference is, that experiences of resistance being those from which the conception of Space is generated, the resistance-attribute of Matter must be regarded as primordial and the space-attribute as derivative. Whence it becomes manifest that our experience of *force*, is that out of which the idea of Matter is built. Matter as opposing our muscular energies, being immediately present to consciousness in terms of force; and its occupancy of Space being known by an abstract of experiences originally given in terms of force; it follows that forces, standing in certain correlations, form the whole content of our idea of Matter.

Such being our cognition of the relative reality, what are we to say of the absolute reality? We can only say that it is some mode of the Unknowable, related to the Matter we know, as cause to effect. The relativity of our cognition of Matter is shown alike by the above analysis, and by the contradictions which are evolved when we deal with the cognition as an absolute one (§ 16). But, as we have lately seen, though known to us only under relation, Matter is as real in the true sense of that word, as it would be could we know it out of relation; and further, the relative reality which we know as Matter, is necessarily represented to the mind as standing in a persistent or real relation to the absolute reality. We may therefore deliver ourselves over without hesitation, to those terms of thought which experience has organized in us. We need not in our physical, chemical, or other researches, refrain from dealing with Matter as made up of extended and resistent atoms; for this conception, necessarily resulting from our experiences of Matter, is not less legitimate, than the conception of aggregate masses as extended and resistent. The atomic hypothesis, as well as the kindred hypothesis of an all-pervading ether consisting of molecules, is simply a necessary development of those universal forms which the actions of the Unknowable have wrought in us. The conclusions logically worked out by the aid of these hypotheses, are sure to be in harmony with all others which these same forms involve, and will have a relative truth that is equally complete.

§ 64. The conception of Motion as presented or represented in the developed consciousness, involves the conceptions of Space, of Time, and of Matter. A something that moves; a series of positions occupied in succession;

and a group of co-existent positions united in thought with the successive ones—these are the constituents of the idea. And since, as we have seen, these are severally elaborated from experiences of *force* as given in certain correlations, it follows that from a further synthesis of such experiences, the idea of Motion is also elaborated. A certain other element in the idea, which is in truth its fundamental element, (namely, the necessity which the moving body is under to go on changing its position), results immediately from the earliest experiences of force. Movements of different parts of the organism in relation to each other, are the first presented in consciousness. These, produced by the action of the muscles, necessitate reactions upon consciousness in the shape of sensations of muscular tension. Consequently, each stretching-out or drawing-in of a limb, is originally known as a series of muscular tensions, varying in intensity as the position of the limb changes. And this rudimentary consciousness of Motion, consisting of serial impressions of force, becomes inseparably united with the consciousness of Space and Time as fast as these are abstracted from further impressions of force. Or rather, out of this primitive conception of Motion, the adult conception of it is developed simultaneously with the development of the conceptions of Space and Time: all three being evolved from the more multiplied and varied impressions of muscular tension and objective resistance. Motion, as we know it, is thus traceable, in common with the other ultimate scientific ideas, to experiences of force.

That this relative reality answers to some absolute reality, it is needful only for form's sake to assert. What has been said above, respecting the Unknown Cause which produces in us the effects called Matter, Space, and Time, will apply, on simply changing the terms, to Motion.

§ 65. We come down then finally to Force, as the ultimate of ultimates. Though Space, Time, Matter, and Motion, are apparently all necessary data of intelligence, yet a psychological analysis (here indicated only in rude outline) shows us that these are either built up of, or abstracted from, experiences of Force. Matter and Motion, as we know them, are differently conditioned manifestations of Force. Space and Time, as we know them, are disclosed along with these different manifestations of Force as the conditions under which they are presented. Matter and Motion are concretes built up from the *contents* of various mental relations; while Space and Time are abstracts of the *forms* of these various relations. Deeper down than these, however, are the primordial experiences of Force, which, as occurring in consciousness in different combinations, supply at once the materials whence the forms of relations are generalized, and the related objects built up. A single impression of force is manifestly receivable by a sentient being devoid of mental forms: grant but sensibility, with no established power of thought, and a force producing some nervous change, will still be presentable at the supposed seat of sensation. Though no single impression of force so received, could itself produce consciousness (which implies relations between different states), yet a multiplication of such impressions, differing in kind and degree, would give the materials for the establishment of relations, that is, of thought. And if such relations differed in their forms as well as in their contents, the impressions of such forms would be organized simultaneously with the impressions they contained. Thus all other modes of consciousness are derivable from experiences of Force; but experiences of Force are not derivable from anything else. Indeed, it needs but to remember that consciousness consists of changes, to see that the ultimate datum of consciousness must be that of which change is the manifestation; and that thus the force by which we ourselves produce changes, and which serves to symbolize the cause of changes in general, is the final disclosure of analysis.

It is a truism to say that the nature of this undecomposable element of our knowledge is inscrutable. If, to use an algebraic illustration, we represent Matter, Motion, and Force, by the symbols x, y, and z; then, we may ascertain the values of x and y in terms of z; but the value of z can never be found: z is the unknown quantity which must for ever remain unknown; for the obvious reason that there is nothing in which its value can be expressed. It is within the possible reach of our intelligence to go on simplifying the equations of all phenomena, until the complex symbols which formulate them are reduced to certain functions of this ultimate symbol; but when we have done this, we have reached that limit which eternally divides science from nescience.

That this undecomposable mode of consciousness into which all other modes may be decomposed, cannot be itself the Power manifested to us through phenomena, has been already proved (§ 18). We saw that to assume an identity of nature between the cause of changes as it absolutely exists, and that cause of change of which we are conscious in our own muscular efforts, betrays us into alternative impossibilities of thought. Force, as we know it, can be regarded only as a certain conditioned effect of the Unconditioned Cause—as the relative reality indicating to us an Absolute Reality by which it is immediately produced. And here, indeed, we see even more clearly than before, how inevitable is that transfigured realism to which sceptical criticism finally brings us round. Getting rid of all complications, and contemplating pure Force, we are irresistibly compelled by the relativity of our thought, to vaguely conceive some unknown force as the correlative of the known force. Conditioned effect and unconditioned cause, are here presented in their primordial relation as two sides of the same change; of which we are obliged to regard the conditioned and the unconditioned sides as equally real: the only difference being that the reality of the one is made relative by the imposition of the forms and limits of our consciousness, while the reality of the other, in the absence of those forms and limits, remains absolute.

Thus much respecting the nature of our ultimate scientific ideas. Before proceeding to our general inquiry concerning the causes of Evolution, we have still to consider certain ultimate scientific truths.

> 11. For the psychological conclusions briefly set forth in this and the three sections following it, the justification will be found in the writer's *Principles of Psychology*.

CHAPTER VI
THE INDESTRUCTIBILITY OF MATTER

§ 66. Not because the truth is unfamiliar, is it needful here to say something concerning the indestructibility of Matter; but partly because the symmetry of our argument demands the enunciation of this truth, and partly because the evidence on which it is accepted requires examination. Could it be shown, or could it with any rationality be even supposed, that Matter, either in its aggregates or in its units, ever became non-existent, there would be an end to the inquiry on which we are now setting out. Evolution, considered as a re-arrangement of parts, could not be scientifically explained, if, during its course, any of the parts might arise out of nothing or might lapse into nothing. The question would no longer be one comprehending only the forces and motions by which the re-arrangement is effected; but would involve an incalculable element, and would hence be insoluble. Clearly, therefore, the indestructibility of Matter is an indispensable axiom.

So far from being admitted as a self-evident truth, this would, in primitive times, have been rejected as a self-evident error. There was once universally current, a notion that things could vanish into absolute nothing, or arise out of absolute nothing. If we analyze early superstitions, or that faith in magic which was general in later times and even still survives among the uncultured, we find one of its postulates to be, that by some potent spell Matter can be called out of nonentity, and can be made non-existent. If men did not believe this in the strict sense of the word (which would imply that the process of creation or annihilation was clearly represented in consciousness), they still believed that they believed it; and how nearly, in their confused thoughts, the one was equivalent to the other, is shown by their conduct. Nor, indeed, have dark ages and inferior minds alone betrayed this belief. The current theology, in its teachings respecting the beginning and end of the world, is clearly pervaded by it; and it may be even questioned whether Shakespeare, in his poetical anticipation of a time when all things should disappear and "leave not a wrack behind," was not under its influence. The gradual accumulation of experiences however, and still more the organization of experiences, has tended slowly to reverse this conviction; until now, the doctrine that Matter is indestructible has become a common-place. Whatever may be true of it absolutely, we have learnt that relatively to our consciousness, Matter never either comes into existence or ceases to exist. Cases which once gave an apparent support to the illusion that something could come out of nothing, a wider knowledge has one by one cancelled. The comet that is all at once discovered in the heavens and nightly waxes larger, is proved not to be a newly-created body, but a body that was until lately beyond the range of vision. The cloud which in the course of a few minutes forms in the sky, consists not of substance that has just begun to be, but of substance that previously existed in a more diffused and transparent form. And similarly with a crystal or precipitate in relation to the fluid depositing it. Conversely, the seeming annihilations of Matter turn out, on closer observation, to be only changes of state. It is found that the evaporated water, though it has become invisible, may be brought by condensation to its original shape. The discharged fowling-piece gives evidence that though the gunpowder has disappeared, there have appeared in place of it certain gases, which, in assuming a larger volume, have caused the explosion. Not, however, until the rise of quantitative chemistry, could the conclusion suggested by such experiences be reduced to a certainty. When, having ascertained not only the combinations into which various substances enter, but also the proportions in which they combine, chemists were enabled to account for the matter that had made its appearance or become invisible, the proof was rendered complete. When, in place of the candle that had slowly burnt away, it was shown that certain calculable quantities of carbonic acid and water had resulted—when it was demonstrated that the joint weight of the carbonic acid and water thus produced, was equal to the weight of the candle plus that of the oxygen uniting with its constituents during combustion; it was put beyond doubt that the carbon and hydrogen forming the candle, were still in existence, and had simply changed their state. And of the general conclusion thus exemplified, the exact analyses daily made, in which the same portion of matter is pursued through numerous transformations and finally separated, furnish never-ceasing confirmations.

Such has become the effect of this specific evidence, joined to that general evidence which the continued existence of familiar objects unceasingly gives us; that the indestructibility of Matter is now recognized by many as a truth of which the negation is inconceivable. Habitual experiences being no longer met by any counter-experiences, as they once seemed to be; but these apparent counter-experiences furnishing new proof that Matter exists permanently, even where the senses fail to detect it; it has grown into an axiom of science, that whatever metamorphoses Matter undergoes, its quantity is fixed. The chemist, the physicist, and the physiologist, not only one and all take this for granted, but would severally profess themselves unable to realize any supposition to the contrary.

§ 67. This last fact naturally raises the question, whether we have any higher warrant for this fundamental belief, than the warrant of conscious induction. The indestructibility of Matter is proved experimentally to be an absolute uniformity within the range of our experience. But absolute uniformities of experience, generate absolute uniformities of thought. Does it not follow, then, that this ultimate truth must be a cognition involved in our mental organization? An affirmative answer we shall find unavoidable.

What is termed the ultimate incompressibility of Matter, is an admitted law of thought. Though it is possible to imagine a piece of matter to be compressed without limit, yet however small the bulk to which we conceive it reduced, it is impossible to conceive it reduced into nothing. While we can represent to ourselves the parts of the matter as indefinitely approximated, and the space occupied as indefinitely decreased, we cannot represent to ourselves the quantity of matter as made less. To do this would imply an imagined disappearance of some of the constituent parts—would imply that some of the constituent parts were in thought compressed into nothing; which is no more possible than the compression of the whole into nothing. Whence it is an obvious corollary, that the total quantity of matter in the Universe, cannot really be conceived as diminished, any more than it can be conceived as increased. Our inability to conceive Matter becoming non-existent, is immediately consequent on the very nature of thought. Thought consists in the establishment of relations. There can be no relation established, and therefore no thought framed, when one of the related terms is absent from consciousness. Hence it is impossible to think of something becoming nothing, for the same reason that it is impossible to think of nothing becoming something—the reason, namely, that nothing cannot become an object of consciousness. The annihilation of Matter is unthinkable for the same reason that the creation of Matter is unthinkable; and its indestructibility thus becomes an *à priori* cognition of the highest order—not one that results from a long continued registry of experiences gradually organized into an irreversible mode of thought; but one that is given in the form of all experiences whatever.

Doubtless it will be considered strange that a truth only in modern times accepted as unquestionable, and then only by men of science, should be classed as an *à priori* truth; not only of equal certainty with those commonly so classed, but of even higher certainty. To set down as a proposition which cannot be thought, one which mankind once universally professed to think, and which the great majority profess to think even now, seems absurd. The explanation is, that in this, as in countless other cases, men have supposed themselves to think what they did not think. As was shown at the outset, the greater part of our conceptions are symbolic. Many of these symbolic conceptions, though rarely developed into real ones, admit of being so developed; and, being directly or indirectly proved to correspond with actualities, are valid. But along with these there pass current others which cannot be developed—cannot by any direct or indirect process be realized in thought; much less proved to correspond with actualities. Not being habitually tested, however, the legitimate and illegitimate symbolic conceptions are confounded together; and supposing themselves to have literally thought, that which they have thought only symbolically, men say they believe propositions of which the terms cannot even be put together in consciousness. Hence the ready acceptance given to sundry hypotheses respecting the origin of the Universe, which yet are absolutely unthinkable. And as before we found the commonly asserted doctrine that Matter was created out of nothing, to have been never really conceived at all, but to have been conceived only symbolically; so here we find the annihilation of Matter to have been conceived only symbolically, and the symbolic conception mistaken for a real one. Possibly it will be objected that the words *thought*, and *belief*, and *conception*, are here employed in new senses; and that it is a misuse of language to say that men did not really think that which has nevertheless so profoundly influenced their conduct. It must be confessed that there is an inconvenience in so restricting the meanings of these words. There is no remedy however. Definite conclusions can be reached, only by the use of well-defined terms. Questions touching the validity of any portion of our knowledge, cannot be profitably discussed unless the words *knowing*, and *thinking*, have specific interpretations. We must not include under them whatever confused processes of consciousness the popular speech applies them to; but only the distinct processes of consciousness. And if this obliges us to reject a large part of human thinking as not thinking at all, but merely pseudo-thinking, there is no help for it.

Returning to the general question, we find the results to be:—that we have positive experience of the continued existence of Matter; that the form of our thought renders it impossible for us to have experience of Matter passing into non-existence, since such experience would involve cognition of a relation having one of its terms not representable in consciousness; that hence the indestructibility of Matter is in strictness an *à priori* truth; that nevertheless, certain illusive experiences, suggesting the notion of its annihilation, have produced in undisciplined minds not only the supposition that Matter could be conceived to become non-existent, but the notion that it did so; but that careful observation, showing the supposed annihilations to have never taken place, has confirmed, *à posteriori*, the *à priori* cognition which Psychology shows to result from a uniformity of experience that can never be met by counter-experience.

§ 68. The fact, however, which it most concerns us here to observe, is, the nature of the perceptions by which the permanence of Matter is perpetually illustrated to us, and from which Science draws the inference that Matter is indestructible. These perceptions, under all their forms, amount simply to this—that the *force* which a given quantity of matter exercises, remains always the same. This is the proof on which common sense and exact science alike rely. When, for example, somebody known to have existed a few years since is said to exist still, by one who yesterday saw him, his assertion amounts to this—that an object which in past time wrought on his consciousness a certain group of changes, still exists because a like group of changes has been again wrought on his consciousness: the continuance of the power thus to impress him, he holds to prove the continuance of the object. Should some auditor allege a mistake in identity, the witness is admitted to give conclusive proof when he says that he not only saw, but shook hands with this person, and remarked while grasping his hand, that absence of the index finger which was his known peculiarity: the implication being, that an object which through a special combination of forces, produces special tactual impressions, is concluded still to exist while it continues still to do this. Even more clearly do we see that force is our ultimate measure of Matter, in those cases where the shape of the matter has been changed. A piece of gold given to an artizan to be worked into an ornament, and which when brought back appears to be less, is placed in the scales; and if it balances a much smaller weight than it did in its rough state, we infer that much has been lost either in manipulation or by direct abstraction. Here the obvious postulate is, that the quantity of Matter is finally determinable by the quantity of gravitative force it manifests. And this is the kind of evidence on which Science bases its experimentally-established induction that Matter is indestructible. Whenever a piece of substance lately visible and tangible, has been reduced to an invisible, intangible shape, but is proved by the weight of the gas into which it has been transformed to be still existing; the assumption is, that though otherwise insensible to us, the amount of matter is the same, if it still tends towards the Earth with the same force. Similarly, every case in which the weight of an element present in combination, is inferred from the known weight of another element which it neutralizes, is a case in which the quantity of matter is expressed in terms of the quantity of chemical force it exerts; and in which this specific chemical force is assumed to be the necessary correlative of a specific gravitative force.

Thus then by the indestructibility of Matter, we really mean the indestructibility of the *force* with which Matter affects us. As we become conscious of Matter only through that resistance which it opposes to our muscular energy, so do we become conscious of the permanence of Matter only through the permanence of this resistance; as either immediately or mediately proved to us. And this truth is made manifest not only by analysis of the *à posteriori* cognition, but equally so by analysis of the *à priori* one. For that which we cannot conceive to be diminished by the continued compression of Matter, is not its occupancy of space, but its ability to resist.

CHAPTER VII
THE CONTINUITY OF MOTION

§ 69. Another general truth of the same order with the foregoing, must here be specified—one which, though not so generally recognized, has yet long been familiar among men of science. The continuity of Motion, like the indestructibility of Matter, is clearly an axiom underlying the very possibility of a rational theory of Evolution. That kind of change in the arrangement of parts, which we have found to constitute Evolution, could not be deductively explained were it possible for Motion either to appear or disappear. If those motions through which the parts pass into a new arrangement, might either proceed from nothing or lapse into nothing, there would be an end to scientific interpretation of them. Each constituent change might as well as not be supposed to begin and end of itself.

The axiomatic character of the truth that Motion is continuous, is recognized only after the discipline of exact science has given precision to the conceptions. Aboriginal men, our uneducated population, and even most of the so-called educated, think in an extremely indefinite manner. From careless observations, they pass by careless reasoning, to conclusions of which they do not contemplate the implications—conclusions which they never develope for the purpose of seeing whether they are consistent. Accepting without criticism the dicta of unaided perception, to the effect that surrounding bodies when put in motion soon return to rest, the great majority tacitly assume that the motion is actually lost. They do not consider whether the phenomenon can be otherwise interpreted; or whether the interpretation they put on it can be mentally realized. They are content with a colligation of mere appearances. But the establishment of certain facts having quite an opposite implication, led to inquiries which have gradually proved such appearances to be illusive. The discovery that the planets revolve round the Sun with undiminishing speed, raised the suspicion that a moving body, when not interfered with, will go on for ever without change of velocity; and suggested the question whether bodies which lose their motion, do not at the same time communicate as much motion to other bodies. It was a familiar fact that a stone would glide further over a smooth surface, such as ice, presenting no small objects to which it could part with its motion by collision, than over a surface strewn with such small objects; and that a projectile would travel a far greater distance through a rare medium like air, than through a dense medium like water. Thus the primitive notion that moving bodies had an inherent tendency gradually to lose their motion and finally stop—a notion of which the Greeks did not get rid, but which lasted till the time of Galileo—began to give way. It was further shaken by such experiments as those of Hooke, which proved that the spinning of a top continued long in proportion as it was prevented from communicating movement to surrounding matter—experiments which, when repeated with the aid of modern appliances, have shown that *in vacuo* such rotation, retarded only by the friction of the axis, will continue for nearly an hour. Thus have been gradually dispersed, the obstacles to the reception of the first law of motion;—the law, namely, that when not influenced by external forces, a moving body will go on in a straight line with a uniform velocity. And this law is in our day being merged in the more general one, that Motion, like Matter, is indestructible; and that whatever is lost by any one portion of matter is transferred to other portions—a conclusion which, however much at variance it seems with cases of sudden arrest from collision with an immovable object, is yet reconciled with such cases by the discovery that the motion apparently lost continues under new forms, though forms not directly perceptible.

§ 70. And here it may be remarked of Motion, as it was before of Matter, that its indestructibility is not only to be inductively inferred, but that it is a necessity of thought: its destructibility never having been truly conceived at all, but having always been, as it is now, a mere verbal proposition that cannot be realized in consciousness—a pseud-idea. Whether that absolute reality which produces in us the consciousness we call Motion, be or be not an eternal mode of the Unknowable, it is impossible for us to say; but that the relative reality which we call Motion never can come into existence, or cease to exist, is a truth involved in the very nature of our consciousness. To think of Motion as either being created or annihilated—to think of nothing becoming something, or something becoming nothing—is to establish in consciousness a relation between two terms of which one is absent from consciousness, which is impossible. The very nature of intelligence, negatives the supposition that Motion can be conceived (much less known) to either commence or cease.

§ 71. It remains to be pointed out that the continuity of Motion, as well as the indestructibility of Matter, is really known to us in terms of *force*. That a certain manifestation of force remains for ever undiminished, is the ultimate content of the thought; whether reached *à posteriori* or *à priori*.

From terrestrial physics let us take the case of sound propagated to a great distance. Whenever we are directly conscious of the causation of sound (namely, when we produce it ourselves), its invariable antecedent is force. The immediate sequence of this force we know to be motion—first, of our own organs, and then of the body which we set vibrating. The vibrations so generated we can discern both through the fingers and through the ears; and that the sensations received by the ears are the equivalents of mechanical force communicated to the air, and by it impressed on surrounding objects, we have clear proof when objects are fractured: as windows by the report of a cannon; or a glass vessel by a powerful voice. On what, then, rests the reasoning when, as occasionally happens under favourable circumstances, men on board a vessel a hundred miles from shore, hear the ringing of church-bells on placing their ears in the focus of the main sail; and when it is inferred that atmospheric undulations have traversed this immense distance? Manifestly, the assertion that the motion of the clapper, transformed into the vibrations of the bell, and communicated to the surrounding air, has propagated itself thus far on all sides, diminishing in intensity as the mass of air moved became greater, is based solely upon a certain change produced in consciousness through the ears. The listeners are not conscious of motion; they are conscious of an impression produced on them—an impression which implies a force as its necessary correlative. With force they begin, and with force they end: the intermediate motion being simply inferred. Again, where, as in celestial physics, the continuity of motion is quantitatively proved, the proof is not direct but inferential; and forces furnish the data for the inference. A particular planet can be identified only by its constant power to affect our visual organs in a special way—to impress upon the retina a group of forces standing in a particular correlation. Further, such planet has not been *seen* to move by the astronomical observer; but its motion is *inferred* from a comparison of its present position with the position it before occupied. If rigorously examined, this comparison proves to be a comparison between the different impressions produced on him by the different adjustments of the observing instruments. Going a step further back, it turns out that this difference is meaningless until shown to correspond with a certain calculated position which the planet must occupy, supposing that no motion has been lost. And if, finally, we examine the implied calculation, we find that it makes allowances for those accelerations and retardations which ellipticity of the orbit involves, as well as those variations of velocity caused by adjacent planets—we find, that is, that the motion is concluded to be indestructible not from the uniform velocity of the planet, but from the constant quantity of motion exhibited when allowance is made for the motion communicated to, or received from, other celestial bodies. And when we ask how this communicated motion is estimated, we discover that the estimate is based upon certain laws of force; which laws, one and all, embody the postulate that force cannot be destroyed. Without the axiom that action and re-action are equal and opposite, astronomy could not make its exact predictions; and we should lack the rigorous inductive proof they furnish that motion can never be lost, but can only be transferred.

Similarly with the *à priori* conclusion that Motion is continuous. That which defies suppression in thought, is really the force which the motion indicates. The unceasing change of position, considered by itself, may be mentally abolished without difficulty. We can readily imagine retardation and stoppage to result from the action of external bodies. But to imagine this, is not possible without an abstraction of the force implied by the motion. We are obliged to conceive this force as impressed in the shape of re-action on the bodies that cause the arrest. And the motion that is communicated to them, we are compelled to regard, not as directly communicated, but as a product of the communicated force. We can mentally diminish the velocity or space-element of motion, by diffusing the momentum or force-element over a larger mass of matter; but the quantity of this force-element, which we regard as the cause of the motion, is unchangeable in thought.

CHAPTER VIII
THE PERSISTENCE OF FORCE [12]

§ 72. Before taking a first step in the rational interpretation of Evolution, it is needful to recognize, not only the facts that Matter is indestructible and Motion continuous, but also the fact that Force persists. An attempt to assign the *causes* of Evolution, would manifestly be absurd, if that agency to which the metamorphosis in general and in detail is due, could either come into existence or cease to exist. The succession of phenomena would in such case be altogether arbitrary; and deductive science impossible.

Here, indeed, the necessity is even more imperative than in the two preceding cases. For the validity of the proofs given that Matter is indestructible and Motion continuous, really depends upon the validity of the proof that Force is persistent. An analysis of the reasoning demonstrated that in both cases, the *à posteriori* conclusion involves the assumption that unchanged quantities of Matter and Motion are proved by unchanged manifestations of Force; and in the *à priori* cognition we found this to be the essential constituent. Hence, that the quantity of Force remains always the same, is the fundamental cognition in the absence of which these derivative cognitions must disappear.

§ 73. But now on what grounds do we assert the persistence of Force? Inductively we can allege no evidence except such as is presented to us throughout the world of sensible phenomena. No force however, save that of which we are conscious during our own muscular efforts, is immediately known to us. All other force is mediately known through the changes we attribute to it. Since, then, we cannot infer the persistence of Force from our own sensation of it, which does not persist; we must infer it, if it is inferred at all, from the continuity of Motion, and the undiminished ability of Matter to produce certain effects. But to reason thus is manifestly to reason in a circle. It is absurd to allege the indestructibility of Matter, because we find experimentally that under whatever changes of form a given mass of matter exhibits the same gravitation, and then afterwards to argue that gravitation is constant because a given mass of matter exhibits always the same quantity of it. We cannot prove the continuity of Motion by assuming that Force is persistent, and then prove the persistence of Force by assuming that Motion is continuous.

The data of both objective and subjective science being involved in this question touching the nature of our cognition that Force is persistent, it will be desirable here to examine it more closely. At the risk of trying the reader's patience, we must reconsider the reasoning through which the indestructibility of Matter and the continuity of Motion are established; that we may see how impossible it is to arrive by parallel reasoning at the persistence of Force. In all three cases the question is one of quantity:—does the Matter, or Motion, or Force, ever diminish in quantity? Quantitative science implies measurement; and measurement implies a unit of measure. The units of measure from which all others of any exactness are derived, are units of linear extension. From these, through the medium of the equal-armed lever or scales, we derive our equal units of weight, or gravitative force. And it is by means of these equal units of extension and equal units of weight, that we make those quantitative comparisons by which the truths of exact science are reached. Throughout the investigations leading the chemist to the conclusion that of the carbon which has disappeared during combustion, no portion has been lost, and that in any compound afterwards formed by the resulting carbonic acid the whole of the original carbon is present, what is his repeatedly assigned proof? That afforded by the scales. In what terms is the verdict of the scales given? In grains—in units of weight—in units of gravitative force. And what is the total content of the verdict? That as many units of gravitative force as the carbon exhibited at first, it exhibits still. The quantity of matter is asserted to be the same, if the number of units of force it counter-balances is the same. The validity of the inference, then, depends entirely upon *the constancy of the units of force*. If the force with which the portion of metal called a grain-weight, tends towards the Earth, has varied, the inference that Matter is indestructible is vicious. Everything turns on the truth of the assumption that the gravitation of the weights is persistent; and of this no proof is assigned, or can be assigned. In the reasonings of the astronomer there is a like implication; from which we may draw the like conclusion. No problem in celestial physics can be solved without the assumption of some unit of force. This unit need not be, like a pound or a ton, one of which we can take direct cognizance. It is requisite only that the mutual attraction which some two of the bodies concerned exercise at a given distance, should be taken as one; so that the other attractions with which the problem deals, may be expressed in terms of this one. Such unit being assumed, the momenta which the respective masses will generate in each other in a given time, are calculated; and compounding these with the momenta they already have, their places at the end of that time are predicted. The prediction is verified by observation. From this, either of two inferences may be drawn. Assuming the masses to be fixed, the motion may be proved to be undiminished; or assuming the motion to be undiminished, the masses may be proved to be fixed. But the validity of one or other inference, depends wholly on the truth of the assumption that the unit of force is unchanged. Let it be supposed that the gravitation of the two bodies towards each other at the given distance, has varied, and the conclusions drawn are no longer true. Nor is it only in their concrete data that the reasonings of terrestrial and celestial physics assume the persistence of Force. They equally assume it in the abstract principle with which they set out; and which they repeat in justification of every step. The equality of action and reaction is taken for granted from beginning to end of either argument; and to assert that action and reaction are equal and opposite, is to assert that Force is persistent. The allegation really amounts to this, that there cannot be an isolated force beginning and ending in nothing; but that any force manifested, implies an equal antecedent force from which it is derived, and against which it is a reaction. Further, that the force so originating cannot disappear without result; but must expend itself in some other manifestation of force, which, in being produced, becomes its reaction; and so on continually. Clearly then the persistence of Force is an ultimate truth of which no inductive proof is possible.

We might indeed be certain, even in the absence of any such analysis as the foregoing, that there must exist some principle which, as being the basis of science, cannot be established by science. All reasoned-out conclusions whatever, must rest on some postulate. As before shown (§ 23), we cannot go on merging derivative truths in those wider and wider truths from which they are derived, without reaching at last a widest truth which can be merged in no other, or derived from no other. And whoever contemplates the relation in which it stands to the truths of science in general, will see that this truth transcending demonstration is the persistence of Force.

§ 74. But now what is the force of which we predicate persistence? It is not the force we are immediately conscious of in our own muscular efforts; for this does not persist. As soon as an outstretched limb is relaxed, the sense of tension disappears. True, we assert that in the stone thrown or in the weight lifted, is exhibited the effect of this muscular tension; and that the force which has ceased to be present in our consciousness, exists elsewhere. But it does not exist elsewhere under any form cognizable by us. It was proved (§ 18), that though, on raising an object from the ground, we are obliged to think of its downward pull as equal and opposite to our upward pull; and though it is impossible to represent these pulls as equal without representing them as like in kind; yet, since their likeness in kind would imply in the object a sensation of muscular tension, which cannot be ascribed to it, we are compelled to admit that force as it exists out of our consciousness, is not force as we know it. Hence the force of which we assert persistence is that Absolute Force of which we are indefinitely conscious as the necessary correlate of the force we know. Thus, by the persistence of Force, we really mean the persistence of some Power which transcends our knowledge and conception. The manifestations, as occurring either in ourselves or outside of us, do not persist; but that which persists is the Unknown Cause of these manifestations. In other words, asserting the persistence of Force, is but another mode of asserting an Unconditioned Reality, without beginning or end.

Thus, quite unexpectedly, we come down once more to that ultimate truth in which, as we saw, Religion and Science coalesce. On examining the data underlying a rational theory of Evolution, we find them all at last resolvable into that datum without which consciousness was shown to be impossible—the continued existence of an Unknowable as the necessary correlative of the Knowable. Once commenced, the analysis of the truths taken for granted in scientific inquiries, inevitably brings us down to this deepest truth, in which Common Sense and Philosophy are reconciled.

The arguments and conclusion contained in this and the foregoing three chapters, supply, indeed, the complement to the arguments and conclusion set forth in the preceding part of this work. It was there first shown, by an examination of our ultimate religious ideas, that knowledge of Absolute Being is impossible; and the impossibility of knowing Absolute Being, was also shown by an examination of our ultimate scientific ideas. In a succeeding chapter a subjective analysis proved, that while, by the very conditions of thought, we are prevented from knowing anything beyond relative being; yet that by these very same conditions of thought, an indefinite consciousness of Absolute Being is necessitated. And here, by objective analysis, we similarly find that the axiomatic truths of physical science, unavoidably postulate Absolute Being as their common basis.

Thus there is even a more profound agreement between Religion and Science than was before shown. Not only are they wholly at one on the negative proposition that the Non-relative cannot be known; but they are wholly at one on the positive proposition that the Non-relative is an actual existence. Both are obliged by the demonstrated untenability of their supposed cognitions, to confess that the Ultimate Reality is incognizable; and yet both are obliged to assert the existence of an Ultimate Reality. Without this, Religion has no subject-matter; and without this, Science, subjective and objective, lacks its indispensable datum. We cannot construct a theory of internal phenomena without postulating Absolute Being; and unless we postulate Absolute Being, or being which persists, we cannot construct a theory of external phenomena.

§ 75. A few words must be added respecting the nature of this fundamental consciousness. Already it has been looked at from several points of view; and here it seems needful finally to sum up the results.

In Chapter IV. we saw that the Unknown Power of which neither beginning nor end can be conceived, is present to us as that unshaped material of consciousness which is shaped afresh in every thought. Our inability to conceive its limitation, is thus simply the obverse of our inability to put an end to the thinking subject while still continuing to think. In the two foregoing chapters, we contemplated this fundamental truth under another aspect. The indestructibility of Matter and the continuity of Motion, we saw to be really corollaries from the impossibility of establishing in thought a relation between something and nothing. What we call the establishment of a relation in thought, is the passage of the substance of consciousness, from one form into another. To think of something becoming nothing, would involve that this substance of consciousness having just existed under a given form, should next assume no form; or should cease to be consciousness. And thus our inability to conceive Matter and Motion destroyed, is our inability to suppress consciousness itself. What, in these two foregoing chapters, was proved true of Matter and Motion, is, *à fortiori*, true of the Force out of which our conceptions of Matter and Motion are built. Indeed, as we saw, that which is indestructible in matter and motion, is the force they present. And, as we here see, the truth that Force is indestructible, is the obverse of the truth that the Unknown Cause of the changes going on in consciousness is indestructible. So that the persistence of consciousness, constitutes at once our immediate experience of the persistence of Force, and imposes on us the necessity we are under of asserting its persistence.

§ 76. Thus, in all ways there is forced on us the fact, that here is an ultimate truth given in our mental constitution. It is not only a datum of science, but it is a datum which even the assertion of our nescience involves. Whoever alleges that the inability to conceive a beginning or end of the Universe, is a *negative* result of our mental structure, cannot deny that our consciousness of the Universe as persistent, is a *positive* result of our mental structure. And this persistence of the Universe, is the persistence of that Unknown Cause, Power, or Force, which is manifested to us through all phenomena.

Such then is the foundation of any possible system of positive knowledge. Deeper than demonstration—deeper even than definite cognition—deep as the very nature of mind, is the postulate at which we have arrived. Its authority transcends all other whatever; for not only is it given in the constitution of our own consciousness, but it is impossible to imagine a consciousness so constituted as not to give it. Thought, involving simply the establishment of relations, may be readily conceived to go on while yet these relations have not been organized into the abstracts we call Space and Time; and so there is a conceivable kind of consciousness which does not contain the truths, commonly called *à priori*, involved in the organization of these forms of relations. But thought cannot be conceived to go on without some element between which its relations may be established; and so there is no conceivable kind of consciousness which does not imply continued existence as its datum. Consciousness without this or that particular *form* is possible; but consciousness without *contents* is impossible.

The sole truth which transcends experience by underlying it, is thus the persistence of Force. This being the basis of experience, must be the basis of any scientific organization of experiences. To this an ultimate analysis brings us down; and on this a rational synthesis must build up.

12. Some two years ago, I expressed to my friend Professor Huxley, my dissatisfaction with the current expression—"Conservation of Force;" assigning as reasons, first, that the word "conservation" implies a conserver and an act of conserving; and, second, that it does not imply the existence of the force before that particular manifestation of it with which we commence. In place of "conservation," Professor Huxley suggested *persistence*. This entirely meets the first of the two objections; and though the second may be urged against it, no other word less faulty in this respect can be found. In the absence of a word specially coined for the purpose, it seems the best; and as such I adopt it.

CHAPTER IX
THE CORRELATION AND
EQUIVALENCE OF FORCES

§ 77. When, to the unaided senses, Science began to add supplementary senses in the shape of measuring instruments, men began to perceive various phenomena which eyes and fingers could not distinguish. Of known forms of force, minuter manifestations became appreciable; and forms of force before unknown were rendered cognizable and measurable. Where forces had apparently ended in nothing, and had been carelessly supposed to have actually done so, instrumental observation proved that effects had in every instance been produced: the forces reappearing in new shapes. Hence there has at length arisen the inquiry whether the force displayed in each surrounding change, does not in the act of expenditure undergo metamorphosis into an equivalent amount of some other force or forces. And to this inquiry experiment is giving an affirmative answer, which becomes day by day more decisive. Grove, Helmholtz, and Meyer, are more than any others to be credited with the clear enunciation of this doctrine. Let us glance at the evidence on which it rests.

Motion, wherever we can directly trace its genesis, we find to pre-exist as some other mode of force. Our own voluntary acts have always certain sensations of muscular tension as their antecedents. When, as in letting fall a relaxed limb, we are conscious of a bodily movement requiring no effort, the explanation is that the effort was exerted in raising the limb to the position whence it fell. In this case, as in the case of an inanimate body descending to the Earth, the force accumulated by the downward motion is just equal to the force previously expended in the act of elevation. Conversely, Motion that is arrested produces, under different circumstances, heat, electricity, magnetism, light. From the warming of the hands by rubbing them together, up to the ignition of a railway-brake by intense friction—from the lighting of detonating powder by percussion, up to the setting on fire a block of wood by a few blows from a steam-hammer; we have abundant instances in which heat arises as Motion ceases. It is uniformly found, that the heat generated is great in proportion as the Motion lost is great; and that to diminish the arrest of motion, by diminishing the friction, is to diminish the quantity of heat evolved. The production of electricity by Motion is illustrated equally in the boy's experiment with rubbed sealing-wax, in the common electrical machine, and in the apparatus for exciting electricity by the escape of steam. Wherever there is friction between heterogeneous bodies, electrical disturbance is one of the consequences. Magnetism may result from Motion either immediately, as through percussion on iron, or mediately as through electric currents previously generated by Motion. And similarly, Motion may create light; either directly, as in the minute incandescent fragments struck off by violent collisions, or indirectly, as through the electric spark. "Lastly, Motion may be again reproduced by the forces which have emanated from Motion; thus, the divergence of the electrometer, the revolution of the electrical wheel, the deflection of the magnetic needle, are, when resulting from frictional electricity, palpable movements reproduced by the intermediate modes of force, which have themselves been originated by motion."

That mode of force which we distinguish as Heat, is now generally regarded by physicists as molecular motion—not motion as displayed in the changed relations of sensible masses to each other, but as occurring among the units of which such sensible masses consist. If we cease to think of Heat as that particular sensation given to us by bodies in certain conditions, and consider the phenomena otherwise presented by these bodies, we find that motion, either in them or in surrounding bodies, or in both, is all that we have evidence of. With one or two exceptions which are obstacles to every theory of Heat, heated bodies expand; and expansion can be interpreted only as a movement of the units of a mass in relation to each other. That so-called radiation through which anything of higher temperature than things around it, communicates Heat to them, is clearly a species of motion. Moreover, the evidence afforded by the thermometer that Heat thus diffuses itself, is simply a movement caused in the mercurial column. And that the molecular motion which we call Heat, may be transformed into visible motion, familiar proof is given by the steam-engine; in which "the piston and all its concomitant masses of matter are moved by the molecular dilatation of the vapour of water." Where Heat is absorbed without apparent result, modern inquiries show that decided though unobtrusive changes are produced: as on glass, the molecular state of which is so far changed by heat, that a polarized ray of light passing through it becomes visible, which it does not do when the glass is cold; or as on polished metallic surfaces, which are so far changed in structure by thermal radiations from objects very close to them, as to retain permanent impressions of such objects. The transformation of Heat into electricity, occurs when dissimilar metals touching each other are heated at the point of contact: electric currents being so induced. Solid, incombustible matter introduced into heated gas, as lime into the oxyhydrogen flame, becomes incandescent; and so exhibits the conversion of Heat into light. The production of magnetism by Heat, if it cannot be proved to take place directly, may be proved to take place indirectly through the medium of electricity. And through the same medium may be established the correlation of Heat and chemical affinity—a correlation which is indeed implied by the marked influence that Heat exercises on chemical composition and decomposition.

The transformations of Electricity into other modes of force, are still more clearly demonstrable. Produced by the motion of heterogeneous bodies in contact, Electricity, through attractions and repulsions, will immediately reproduce motion in neighbouring bodies. Now a current of Electricity generates magnetism in a bar of soft iron; and now the rotation of a permanent magnet generates currents of Electricity. Here we have a battery in which from the play of chemical affinities an electric current results; and there, in the adjacent cell, we have an electric current effecting chemical decomposition. In the conducting wire we witness the transformation of Electricity into heat; while in electric sparks and in the voltaic arc we see light produced. Atomic arrangement, too, is changed by Electricity: as instance the transfer of matter from pole to pole of a battery; the fractures caused by the disruptive discharge; the formation of crystals under the influence of electric currents. And whether, conversely, Electricity be or be not directly generated by re-arrangement of the atoms of matter, it is at any rate indirectly so generated through the intermediation of magnetism.

How from Magnetism the other physical forces result, must be next briefly noted—briefly, because in each successive case the illustrations become in great part the obverse forms of those before given. That Magnetism produces motion is the ordinary evidence we have of its existence. In the magneto-electric machine we see a rotating magnet evolving electricity. And the electricity so evolved may immediately after exhibit itself as heat, light, or chemical affinity. Faraday's discovery of the effect of Magnetism on polarized light, as well as the discovery that change of magnetic state is accompanied by heat, point to further like connexions. Lastly, various experiments show that the magnetization of a body alters its internal structure; and that conversely, the alteration of its internal structure, as by mechanical strain, alters its magnetic condition.

Improbable as it seemed, it is now proved that from Light also may proceed the like variety of agencies. The solar rays change the atomic arrangements of particular crystals. Certain mixed gases, which do not otherwise combine, combine in the sunshine. In some compounds Light produces decomposition. Since the inquiries of photographers have drawn attention to the subject, it has been shown that "a vast number of substances, both elementary and compound, are notably affected by this agent, even those apparently the most unalterable in character, such as metals." And when a daguerreotype plate is connected with a proper apparatus "we get chemical action on the plate, electricity circulating through the wires, magnetism in the coil, heat in the helix, and motion in the needles."

The genesis of all other modes of force from Chemical Action, scarcely needs pointing out. The ordinary accompaniment of chemical combination is heat; and when the affinities are intense, light also is, under fit conditions, produced. Chemical changes involving alteration of bulk, cause motion, both in the combining elements and in adjacent masses of matter: witness the propulsion of a bullet by the explosion of gunpowder. In the galvanic battery we see electricity resulting from chemical composition and decomposition. While through the medium of this electricity, Chemical Action produces magnetism.

These facts, the larger part of which are culled from Mr. Grove's work on "The Correlation of Physical Forces," show us that each force is transformable, directly or indirectly, into the others. In every change Force undergoes metamorphosis; and from the new form or forms it assumes, may subsequently result either the previous one or any of the rest, in endless variety of order and combination. It is further becoming manifest that the physical forces stand not simply in qualitative correlations with each other, but also in quantitative correlations. Besides proving that one mode of force may be transformed into another mode, experiments illustrate the truth that from a definite amount of one, definite amounts of others always arise. Ordinarily it is indeed difficult to show this; since it mostly happens that the transformation of any force is not into some one of the rest but into several of them: the proportions being determined by the ever-varying conditions. But in certain cases, positive results have been reached. Mr. Joule has ascertained that the fall of 772 lbs. through one foot, will raise the temperature of a pound of water one degree of Fahrenheit. The investigations of Dulong, Petit and Neumann, have proved a relation in amount between the affinities of combining bodies and the heat evolved during their combination. Between chemical action and voltaic electricity, a quantitative connexion has also been established: Faraday's experiments implying that a specific measure of electricity is disengaged by a given measure of chemical action. The well-determined relations between the quantities of heat generated and water turned into steam, or still better the known expansion produced in steam by each additional degree of heat, may be cited in further evidence. Whence it is no longer doubted that among the several forms which force assumes, the quantitative relations are fixed.

The conclusion tacitly agreed on by physicists, is, not only that the physical forces undergo metamorphoses, but that a certain amount of each is the constant equivalent of certain amounts of the others.

§ 78. Throughout Evolution under all its phases, this truth of course invariably holds. Every successive change or group of changes forming part of it, is of necessity limited by the conditions thus implied. The forces which any step in Evolution exhibits, must be affiliable on the like or unlike forces previously existing; while from the forces so generated must thereafter be derived others more or less transformed. And besides recognizing the forces at any time existing, as necessarily linked with those preceding and succeeding them, we must also recognize the amounts of these forces successively manifested as determinate,—as necessarily producing such and such quantities of results, and as necessarily limited to those quantities.

Involved as are the phenomena of Evolution, it is not to be expected that a *definite* quantitative relation can in each case, or indeed in any case, be shown between the forces expended in successive phases. We have not adequate data for this; and probably shall never have them. The antecedents of the simpler forms of Evolution, belong to a remote past respecting which we can have nothing but inferential knowledge; while the antecedents of the only kind of Evolution which is traceable from beginning to end (namely, that of individual organisms) are too complex to be dealt with by exact methods. Hence we cannot hope to establish *equivalence* among the successive manifestations of force which each order of Evolution affords. The most we can hope is to establish a qualitative correlation that is indefinitely quantitative—quantitative in so far as involving something like a due proportion between causes and effects. If this can be done, however, some progress will be made towards the solution of our problem. Though it may be beyond our power to show a measurable relation between the force or group of forces which any phase of Evolution displays, and the force or group of forces immediately succeeding it; yet if we can show that there always are antecedent forces, and that the effects they produce always become the antecedents of further ones—if while unable to calculate how much of each change will be produced, we can prove that a change of that kind was necessitated—if we can discern even the vaguest correspondence between the amount of such change and the amount of the pre-existing force; we shall advance a step towards interpreting the transformation of the simple into the complex.

With the view of attempting this, let us now reconsider the different types of Evolution awhile since delineated: taking them in the same order as before.

§ 79. On contemplating our Solar System the first fact which strikes us, is, that all its members are in motion; and that their motion is of a two-fold, or rather of a three-fold, kind. Each planet and satellite has a movement of rotation and a movement of translation; besides the movement through space which all have in common with their rotating primary. Whence this unceasing change of place?

The hypothesis of Evolution supplies us with an answer. Impossible as it is to assign a reason for the pre-existence of matter in the diffused form supposed; yet assuming its pre-existence in that form, we have in the gravitation of its parts a cause of motion adequate to the results. So far too as the evidence carries us, we can perceive some quantitative relation between the motions produced, and the gravitative forces expended in producing them. The planets formed from that matter which has travelled the shortest distance towards the common centre of gravity, have the smallest velocities: the uniform law being that in advancing from the outermost to the innermost planets, the rate of orbital motion progressively increases. It may indeed be remarked that this is explicable on the teleological hypothesis; since it is a condition to equilibrium. But without dwelling on the fact that this is beside the question, it will suffice to point out that the like cannot be said of the planetary rotations. No such final cause can be assigned for the rapid axial movement of Jupiter and Saturn, or the slow axial movement of Mercury. But if in pursuance of the doctrine of correlation we look for the antecedents of these gyrations which all planets exhibit, the theory of Evolution furnishes us with equivalent ones; and ones which bear manifest quantitative relations to the motions displayed. For the planets that turn on their axes with extreme rapidity, are those having great masses and large orbits—those, that is, of which the once diffused elements moved to their centres of gravity through immense spaces, and so acquired high velocities. While, conversely, there has resulted the smallest axial movement where the orbit and the mass are both the smallest.

"But what," it may be asked, "has in such case become of all that motion which brought about the aggregation of this diffused matter into solid bodies?" The rotation of each body can be but a residuary result of concentration—a result due to the imperfect balancing of gravitative movements from opposite points towards the common centre. Such gravitative movements from opposite points must in great measure destroy each other. What then has become of these mutually-destroyed motions? The answer which the doctrine of correlation suggests is—they must have been radiated in the form of heat and light. And this answer the evidence, so far as it goes, confirms. Apart from any speculation respecting the genesis of the solar system, the inquiries of geologists lead to the conclusion that the heat of the Earth's still molten nucleus is but a remnant of the heat which once made molten the entire Earth. The mountainous surfaces of the Moon and of Venus (which alone are near enough to be scrutinized), indicating, as they do, crusts that have, like our own, been corrugated by contraction, imply that these bodies too have undergone refrigeration—imply in each of them a primitive heat, such as the hypothesis necessitates. Lastly, we have in the Sun a still-continued production of this heat and light, which must result from the arrest of diffused matter moving towards a common centre of gravity. Here also, as before, a quantitative relation is traceable. Among the bodies which make up the Solar System, those containing comparatively small amounts of matter whose centripetal motion has been destroyed, have already lost nearly all the produced heat: a result which their relatively larger surfaces have facilitated. But the Sun, a thousand times as great in mass as the largest planet, and having therefore to give off an enormously greater quantity of heat and light due to arrest of moving matter, is still radiating with great intensity.

Thus we see that when, in pursuance of the doctrine of correlation, we ask whence come the forces which our Solar System displays, the hypothesis of Evolution gives us a proximate explanation. If the Solar System once existed in a state of indefinite, incoherent homogeneity, and has progressed to its present state of definite, coherent heterogeneity; then the Motion, Heat, and Light now exhibited by its members, are interpretable as the correlatives of pre-existing forces; and between them and their antecedents we may discern relations that are not only qualitative, but also rudely quantitative. How matter came to exist under the form assumed, is a mystery which we must regard as ultimate. But grant such a previous form of existence, and the hypothesis of Evolution interpreted by the laws of correlation, explains for us the forces as we now see them.

§ 80. If we inquire the origin of those forces which have wrought the surface of our planet into its present shape, we find them traceable to the same primordial source as that just assigned. Assuming the solar system to have been evolved, then geologic changes are either direct or indirect results of the unexpended heat caused by nebular condensation. These changes are commonly divided into igneous and aqueous:—heads under which we may most conveniently consider them.

All those periodic disturbances which we call earthquakes, all those elevations and subsidences which they severally produce, all those accumulated effects of many such elevations and subsidences exhibited in ocean-basins, islands, continents, table-lands, mountain-chains, and all those formations which are distinguished as volcanic, geologists now regard as modifications of the Earth's crust produced by the still-molten matter occupying its interior. However untenable may be the details of M. Elie de Beaumont's theory, there is good reason to accept the general proposition that the disruptions and variations of level which take place at intervals on the terrestrial surface, are due to the progressive collapse of the Earth's solid envelope upon its cooling and contracting nucleus. Even supposing that volcanic eruptions, extrusions of igneous rock, and upheaved mountain-chains, could be otherwise satisfactorily accounted for, which they cannot; it would be impossible otherwise to account for those wide-spread elevations and depressions whence continents and oceans result. The conclusion to be drawn is, then, that the forces displayed in these so-called igneous changes, are derived positively or negatively from the unexpended heat of the Earth's interior. Such phenomena as the fusion or agglutination of sedimentary deposits, the warming of springs, the sublimation of metals into the fissures where we find them as ores, may be regarded as positive results of this residuary heat; while fractures of strata and alterations of level are its negative results, since they ensue on its escape. The original cause of all these effects is still, however, as it has been from the first, the gravitating movement of the Earth's matter towards the Earth's centre; seeing that to this is due both the internal heat itself and the collapse which takes place as it is radiated into space.

When we inquire under what forms previously existed the force which works out the geological changes classed as aqueous, the answer is less obvious. The effects of rain, of rivers, of winds, of waves, of marine currents, do not manifestly proceed from one general source. Analysis, nevertheless, proves to us that they have a common genesis. If we ask,—Whence comes the power of the river-current, bearing sediment down to the sea? the reply is,—The gravitation of water throughout the tract which this river drains. If we ask,—How came the water to be dispersed over this tract? the reply is,—It fell in the shape of rain. If we ask,—How came the rain to be in that position whence it fell? the reply is,—The vapour from which it was condensed was drifted there by the winds. If we ask,—How came this vapour to be at that elevation? the reply is,—It was raised by evaporation. And if we ask,—What force thus raised it? the reply is,—The sun's heat. Just that amount of gravitative force which the sun's heat overcame in raising the atoms of water, is given out again in the fall of those atoms to the same level. Hence the denudations effected by rain and rivers, during the descent of this condensed vapour to the level of the sea, are indirectly due to the sun's heat. Similarly with the winds that transport the vapours hither and thither. Consequent as atmospheric currents are on differences of temperature (either general, as between the equatorial and polar regions, or special as between tracts of the Earth's surface of unlike physical characters) all such currents are due to that source from which the varying quantities of heat proceed. And if the winds thus originate, so too do the waves raised by them on the sea's surface. Whence it follows that whatever changes waves produce—the wearing away of shores, the breaking down of rocks into shingle, sand, and mud—are also traceable to the solar rays as their primary cause. The same may be said of ocean-currents. Generated as the larger ones are by the excess of heat which the ocean in tropical climates continually acquires from the Sun; and generated as the smaller ones are by minor local differences in the quantities of solar heat absorbed; it follows that the distribution of sediment and other geological processes which these marine currents effect, are affiliable upon the force which the sun radiates. The only aqueous agency otherwise originating is that of the tides—an agency which, equally with the others, is traceable to unexpended astronomical motion. But making allowance for the changes which this works, we reach the conclusion that the slow wearing down of continents and gradual filling up of seas, by rain, rivers, winds, waves, and ocean-streams, are the indirect effects of solar heat.

Thus the implication forced on us by the doctrine of correlation, that the forces which have moulded and re-moulded the Earth's crust must have pre-existed under some other shape, is quite in conformity with the theory of Evolution; since this pre-supposes certain forces that are both adequate to the results, and cannot be expended without producing the results. We see that while the geological changes classed as igneous, result from the still-progressing motion of the Earth's substance to its centre of gravity; the antagonistic changes classed as aqueous, result from the still-progressing motion of the Sun's substance towards its centre of gravity—a motion which, transformed into heat and radiated to us, is here re-transformed, directly into motions of the gaseous and liquid matters on the Earth's surface, and indirectly into motions of the solid matters.

§ 81. That the forces exhibited in vital actions, vegetal and animal, are similarly derived, is so obvious a deduction from the facts of organic chemistry, that it will meet with ready acceptance from readers acquainted with these facts. Let us note first the physiological generalizations; and then the generalizations which they necessitate.

Plant-life is all directly or indirectly dependant on the heat and light of the sun—directly dependant in the immense majority of plants, and indirectly dependant in plants which, as the fungi, flourish in the dark: since these, growing as they do at the expense of decaying organic matter, mediately draw their forces from the same original source. Each plant owes the carbon and hydrogen of which it mainly consists, to the carbonic acid and water contained in the surrounding air and earth. The carbonic acid and water must, however, be decomposed before their carbon and hydrogen can be assimilated. To overcome the powerful affinities which hold their elements together, requires the expenditure of force; and this force is supplied by the Sun. In what manner the decomposition is effected we do not know. But we know that when, under fit conditions, plants are exposed to the Sun's rays, they give off oxygen and accumulate carbon and hydrogen. In darkness this process ceases. It ceases too when the quantities of light and heat received are greatly reduced, as in winter. Conversely, it is active when the light and heat are great, as in summer. And the like relation is seen in the fact that while plant-life is luxuriant in the tropics, it diminishes in temperate regions, and disappears as we approach the poles. Thus the irresistible inference is, that the forces by which plants abstract the materials of their tissues from surrounding inorganic compounds—the forces by which they grow and carry on their functions, are forces that previously existed as solar radiations.

That animal life is immediately or mediately dependant on vegetal life is a familiar truth; and that, in the main, the processes of animal life are opposite to those of vegetal life is a truth long current among men of science. Chemically considered, vegetal life is chiefly a process of de-oxidation, and animal life chiefly a process of oxidation: chiefly, we must say, because in so far as plants are expenders of force for the purposes of organization, they are oxidizers (as is shown by the exhalation of carbonic acid during the night); and animals, in some of their minor processes, are probably de-oxidizers. But with this qualification, the general truth is that while the plant, decomposing carbonic acid and water and liberating oxygen, builds up the detained carbon and hydrogen (along with a little nitrogen and small quantities of other elements elsewhere obtained) into branches, leaves, and seeds; the animal, consuming these branches, leaves, and seeds, and absorbing oxygen, recomposes carbonic acid and water, together with certain nitrogenous compounds in minor amounts. And while the decomposition effected by the plant, is at the expense of certain forces emanating from the sun, which are employed in overcoming the affinities of carbon and hydrogen for the oxygen united with them; the recomposition effected by the animal, is at the profit of these forces, which are liberated during the combination of such elements. Thus the movements, internal and external, of the animal, are re-appearances in new forms of a power absorbed by the plant under the shape of light and heat. Just as, in the manner above explained, the solar forces expended in raising vapour from the sea's surface, are given out again in the fall of rain and rivers to the same level, and in the accompanying transfer of solid matters; so, the solar forces that in the plant raised certain chemical elements to a condition of unstable equilibrium, are given out again in the actions of the animal during the fall of these elements to a condition of stable equilibrium.

Besides thus tracing a qualitative correlation between these two great orders of organic activity, as well as between both of them and inorganic agencies, we may rudely trace a quantitative correlation. Where vegetal life is abundant, we usually find abundant animal life; and as we advance from torrid to temperate and frigid climates, the two decrease together. Speaking generally, the animals of each class reach a larger size in regions where vegetation is abundant, than in those where it is sparse. And further, there is a tolerably apparent connexion between the quantity of energy which each species of animal expends, and the quantity of force which the nutriment it absorbs gives out during oxidation.

Certain phenomena of development in both plants and animals, illustrate still more directly the ultimate truth enunciated. Pursuing the suggestion made by Mr. Grove, in the first edition of his work on the "Correlation of the Physical Forces," that a connexion probably exists between the forces classed as vital and those classed as physical, Dr. Carpenter has pointed out that such a connexion is clearly exhibited during incubation. The transformation of the unorganized contents of an egg into the organized chick, is altogether a question of heat: withhold heat and the process does not commence; supply heat and it goes on while the temperature is maintained, but ceases when the egg is allowed to cool. The developmental changes can be completed only by keeping the temperature with tolerable constancy at a definite height for a definite time; that is—only by supplying a definite quantity of heat. In the metamorphoses of insects we may discern parallel facts. Experiments show not only that the hatching of their eggs is determined by temperature, but also that the evolution of the pupa into the imago is similarly determined; and may be immensely accelerated or retarded according as heat is artificially supplied or withheld. It will suffice just to add that the germination of plants presents like relations of cause and effect—relations so similar that detail is superfluous.

Thus then the various changes exhibited to us by the organic creation, whether considered as a whole, or in its two great divisions, or in its individual members, conform, so far as we can ascertain, to the law of correlation. Where, as in the transformation of an egg into a chick, we can investigate the phenomena apart from all complications, we find that the re-arrangement of parts which constitutes evolution, involves expenditure of a pre-existing force. Where it is not, as in the egg or the chrysalis, merely the change of a fixed quantity of matter into a new shape, but where, as in the growing plant or animal, we have an incorporation of matter existing outside, there is still a pre-existing external force at the cost of which this incorporation is effected. And where, as in the higher division of organisms, there remain over and above the forces expended in organization, certain surplus forces expended in movement, these too are indirectly derived from this same pre-existing external force.

§ 82. Even after all that has been said in the foregoing part of this work, many will be alarmed by the assertion, that the forces which we distinguish as mental, come within the same generalization. Yet there is no alternative but to make this assertion: the facts which justify, or rather which necessitate it, being abundant and conspicuous. They fall into the following groups.

All impressions from moment to moment made on our organs of sense, stand in direct correlation with physical forces existing externally. The modes of consciousness called pressure, motion, sound, light, heat, are effects produced in us by agencies which, as otherwise expended, crush or fracture pieces of matter, generate vibrations in surrounding objects, cause chemical combinations, and reduce substances from a solid to a liquid form. Hence if we regard the changes of relative position, of aggregation, or of chemical state, thus arising, as being transformed manifestations of the agencies from which they arise; so must we regard the sensations which such agencies produce in us, as new forms of the forces producing them. Any hesitation to admit that, between the physical forces and the sensations there exists a correlation like that between the physical forces themselves, must disappear on remembering how the one correlation, like the other, is not qualitative only but quantitative. Masses of matter which, by scales or dynamometer, are shown to differ greatly in weight, differ as greatly in the feelings of pressure they produce on our bodies. In arresting moving objects, the strains we are conscious of are proportionate to the momenta of such objects as otherwise measured. Under like conditions the impressions of sounds given to us by vibrating strings, bells, or columns of air, are found to vary in strength with the amount of force applied. Fluids or solids proved to be markedly contrasted in temperature by the different degrees of expansion they produce in the mercurial column, produce in us correspondingly different degrees of the sensation of heat. And similarly unlike intensities in our impressions of light, answer to unlike effects as measured by photometers.

Besides the correlation and equivalence between external physical forces, and the mental forces generated by them in us under the form of sensations, there is a correlation and equivalence between sensations and those physical forces which, in the shape of bodily actions, result from them. The feelings we distinguish as light, heat, sound, odour, taste, pressure, &.c, do not die away without immediate results; but are invariably followed by other manifestations of force. In addition to the excitements of secreting organs, that are in some cases traceable, there arises a contraction of the involuntary muscles, or of the voluntary muscles, or of both. Sensations increase the action of the heart—slightly when they are slight; markedly when they are marked; and recent physiological inquiries imply not only that contraction of the heart is excited by every sensation, but also that the muscular fibres throughout the whole, vascular system, are at the same time more or less contracted. The respiratory muscles, too, are stimulated into greater activity by sensations. The rate of breathing is visibly and audibly augmented both by pleasurable and painful impressions on the nerves, when these reach any intensity. It has even of late been shown that inspiration becomes more frequent on transition from darkness into sunshine,—a result probably due to the increased amount of direct and indirect nervous stimulation involved. When the quantity of sensation is great, it generates contractions of the voluntary muscles, as well as of the involuntary ones. Unusual excitement of the nerves of touch, as by tickling, is followed by almost incontrollable movements of the limbs. Violent pains cause violent struggles. The start that succeeds a loud sound, the wry face produced by the taste of anything extremely disagreeable, the jerk with which the hand or foot is snatched out of water that is very hot, are instances of the transformation of feeling: into motion; and in these cases, as in all others, it is manifest that the quantity of bodily action is proportionate to the quantity of sensation. Even where from pride there is a suppression of the screams and groans expressive of great pain (also indirect results of muscular contraction), we may still see in the clenching of the hands, the knitting of the brows, and the setting of the teeth, that the bodily actions developed are as great, though less obtrusive in their results. If we take emotions instead of sensations, we find the correlation and equivalence equally manifest. Not only are the modes of consciousness directly produced in us by physical forces, re-transformable into physical forces under the form of muscular motions and the changes they initiate; but the like is true of those modes of consciousness which are not directly produced in us by the physical forces. Emotions of moderate intensity, like sensations of moderate intensity, generate little beyond excitement of the heart and vascular system, joined sometimes with increased action of glandular organs. But as the emotions rise in strength, the muscles of the face, body, and limbs, begin to move. Of examples may be mentioned the frowns, dilated nostrils, and stampings of anger; the contracted brows, and wrung hands, of grief; the smiles and leaps of joy; and the frantic struggles of terror or despair. Passing over certain apparent, but only apparent, exceptions, we see that whatever be the kind of emotion, there is a manifest relation between its amount, and the amount of muscular action induced: alike from the erect carriage and elastic step of exhilaration, up to the dancings of immense delight, and from the fidgetiness of impatience up to the almost convulsive movements accompanying great mental agony. To these several orders of evidence must be joined the further one, that between our feelings and those voluntary motions into which they are transformed, there comes the sensation of muscular tension, standing in manifest correlation with both—a correlation that is distinctly quantitative: the sense of strain varying, other things equal, directly as the quantity of momentum generated.

"But how," it may be asked, "can we interpret by the law of correlation the genesis of those thoughts and feelings which, instead of following external stimuli, arise spontaneously? Between the indignation caused by an insult, and the loud sounds or violent acts that follow, the alleged connexion may hold; but whence come the crowd of ideas and the mass of feelings that expend themselves in these demonstrations? They are clearly not equivalents of the sensations produced by the words on the ears; for the same words otherwise arranged, would not have caused them. The thing said bears to the mental action it excites, much the same relation that the pulling of a trigger bears to the subsequent explosion—does not produce the power, but merely liberates it. Whence then arises this immense amount of nervous energy which a whisper or a glance may call forth?" The reply is, that the immediate correlates of these and other such modes of consciousness, are not to be found in the agencies acting on us externally, but in certain internal agencies. The forces called vital, which we have seen to be correlates of the forces called physical, are the immediate sources of these thoughts and feelings; and are expended in producing them. The proofs of this are various. Here are some of them. It is a conspicuous fact that mental action is contingent on the presence of a certain nervous apparatus; and that, greatly obscured as it is by numerous and involved conditions, a general relation may be traced between the size of this apparatus and the quantity of mental action as measured by its results. Further, this apparatus has a particular chemical constitution on which its activity depends; and there is one element in it between the amount of which and the amount of function performed, there is an ascertained connexion: the proportion of phosphorus present in the brain being the smallest in infancy, old age and idiotcy, and the greatest during the prime of life. Note next, that the evolution of thought and emotion varies, other things equal, with the supply of blood to the brain. On the one hand, a cessation of the cerebral circulation, from arrest of the heart's action, immediately entails unconsciousness. On the other hand, excess of cerebral circulation (unless it is such as to cause undue pressure) results in an excitement rising finally to delirium. Not the quantity only, but also the condition of the blood passing through the nervous system, influences the mental manifestations. The arterial currents must be duly aerated, to produce the normal amount of cerebration. At the one extreme, we find that if the blood is not allowed to exchange its carbonic acid for oxygen, there results asphyxia, with its accompanying stoppage of ideas and feelings. While at the other extreme, we find that by the inspiration of nitrous oxide, there is produced an excessive, and indeed irrepressible, nervous activity. Besides the connexion between the development of the mental forces and the presence of sufficient oxygen in the cerebral arteries, there is a kindred connexion between the development of the mental forces and the presence in the cerebral arteries of certain other elements. There must be supplied special materials for the nutrition of the nervous centres, as well as for their oxidation. And how what we may call the quantity of consciousness, is, other things equal, determined by the constituents of the blood, is unmistakably seen in the exaltation that follows when certain chemical compounds, as alcohol and the vegeto-alkalies, are added to it. The gentle exhilaration which tea and coffee create, is familiar to all; and though the gorgeous imaginations and intense feelings of happiness produced by opium and hashish, have been experienced by few, (in this country at least,) the testimony of those who have experienced them is sufficiently conclusive. Yet another proof that the genesis of the mental energies is immediately dependent on chemical change, is afforded by the fact, that the effete products separated from the blood by the kidneys, vary in character with the amount of cerebral action. Excessive activity of mind is habitually accompanied by the excretion of an unusual quantity of the alkaline phosphates. Conditions of abnormal nervous excitement bring on analogous effects. And the "peculiar odour of the insane," implying as it does morbid products in the perspiration, shows a connexion between insanity and a special composition of the circulating fluids—a composition which, whether regarded as cause or consequence, equally implies correlation of the mental and the physical forces. Lastly we have to note that this correlation too, is, so far as we can trace it, quantitative. Provided the conditions to nervous action are not infringed on, and the concomitants are the same, there is a tolerably constant ratio between the amounts of the antecedents and consequents. Within the implied limits, nervous stimulants and anæsthetics produce effects on the thoughts and feelings, proportionate to the quantities administered. And conversely, where the thoughts and feelings form the initial term of the relation, the degree of reaction on the bodily energies is great, in proportion as they are great: reaching in extreme cases a total prostration of physique.

Various classes of facts thus unite to prove that the law of metamorphosis, which holds among the physical forces, holds equally between them and the mental forces. Those modes of the Unknowable which we call motion, heat, light, chemical affinity, &c., are alike transformable into each other, and into those modes of the Unknowable which we distinguish as sensation, emotion, thought: these, in their turns, being directly or indirectly re-transformable into the original shapes. That no idea or feeling arises, save as a result of some physical force expended in producing it, is fast becoming a common place of science; and whoever duly weighs the evidence will see, that nothing but an overwhelming bias in favour of a pre-conceived theory, can explain its non-acceptance. How this metamorphosis takes place—how a force existing as motion, heat, or light, can become a mode of consciousness—how it is possible for aerial vibrations to generate the sensation we call sound, or for the forces liberated by chemical changes in the brain to give rise to emotion—these are mysteries which it is impossible to fathom. But they are not profounder mysteries than the transformations of the physical forces into each other. They are not more completely beyond our comprehension than the natures of Mind and Matter. They have simply the same insolubility as all other ultimate questions. We can learn nothing more than that here is one of the uniformities in the order of phenomena.

§ 83. Of course if the law of correlation and equivalence holds of the forces we class as vital and mental, it must hold also of those which we class as social. Whatever takes place in a society is due to organic or inorganic agencies, or to a combination of the two—results either from the undirected physical forces around, from these physical forces as directed by men, or from the forces of the men themselves. No change can occur in its organization, its modes of activity, or the effects it produces on the face of the Earth, but what proceeds, mediately or immediately, from these. Let us consider first the correlation between the phenomena which societies display, and the vital phenomena.

Social power and life varies, other things equal, with the population. Though different races, differing widely in their fitness for combination, show us that the forces manifested in a society are not necessarily proportionate to the number of people; yet we see that under given conditions, the forces manifested are confined within the limits which the number of people imposes. A small society, no matter how superior the character of its members, cannot exhibit the same quantity of social action as a large one. The production and distribution of commodities must be on a comparatively small scale. A multitudinous press, a prolific literature, or a massive political agitation, is not possible. And there can be but a small total of results in the shape of art-products and scientific discoveries. The correlation of the social with the physical forces through the intermediation of the vital ones, is, however, most clearly shown in the different amounts of activity displayed by the same society according as its members are supplied with different amounts of force from the external world. In the effects of good and bad harvests, we yearly see this relation illustrated. A greatly deficient yield of wheat is soon followed by a diminution of business. Factories are worked half-time, or close entirely; railway traffic falls; retailers find their sales much lessened; house-building is almost suspended; and if the scarcity rises to famine, a thinning of the population still more diminishes the industrial vivacity. Conversely, an unusually abundant harvest, occurring under conditions not otherwise unfavourable, both excites the old producing and distributing agencies and sets up new ones. The surplus social energy finds vent in speculative enterprises. Capital seeking investment carries out inventions that have been lying unutilized. Labour is expended in opening new channels of communication. There is increased encouragement to those who furnish the luxuries of life and minister to the æsthetic faculties. There are more marriages, and a greater rate of increase in population. Thus the social organism grows larger, more complex, and more active. When, as happens with most civilized nations, the whole of the materials for subsistence are not drawn from the area inhabited, but are partly imported, the people are still supported by certain harvests elsewhere grown at the expense of certain physical forces. Our own cotton-spinners and weavers supply the most conspicuous instance of a section in one nation living, in great part, on imported commodities, purchased by the labour they expend on other imported commodities. But though the social activities of Lancashire are due chiefly to materials not drawn from our own soil, they are none the less evolved from physical forces elsewhere stored up in fit forms and then brought here.

If we ask whence come these physical forces from which, through the intermediation of the vital forces, the social forces arise, the reply is of course as heretofore—the solar radiations. Based as the life of a society is on animal and vegetal products; and dependent as these animal and vegetal products are on the light and heat of the sun; it follows that the changes going on in societies are effects of forces having a common origin with those which produce all the other orders of changes that have been analyzed. Not only is the force expended by the horse harnessed to the plough, and by the labourer guiding it, derived from the same reservoir as is the force of the falling cataract and the roaring hurricane; but to this same reservoir are eventually traceable those subtler and more complex manifestations of force which humanity, as socially embodied, evolves. The assertion is a startling one, and by many will be thought ludicrous; but it is an unavoidable deduction which cannot here be passed over.

Of the physical forces that are directly transformed into social ones, the like is to be said. Currents of air and water, which before the use of steam were the only agencies brought in aid of muscular effort for the performance of industrial processes, are, as we have seen, generated by the heat of the sun. And the inanimate power that now, to so vast an extent, supplements human labour, is similarly derived. The late George Stephenson was one of the first to recognize the fact that the force impelling his locomotive, originally emanated from the sun. Step by step we go back—from the motion of the piston to the evaporation of the water; thence to the heat evolved during the oxidation of coal; thence to the assimilation of carbon by the plants of whose imbedded remains coal consists; thence to the carbonic acid from which their carbon was obtained; and thence to the rays of light that de-oxidized this carbonic acid. Solar forces millions of years ago expended on the Earth's vegetation, and since locked up beneath its surface, now smelt the metals required for our machines, turn the lathes by which the machines are shaped, work them when put together, and distribute the fabrics they produce. And in so far as economy of labour makes possible the support of a larger population; gives a surplus of human power that would else be absorbed in manual occupations; and so facilitates the development of higher kinds of activity; it is clear that these social forces which are directly correlated with physical forces anciently derived from the sun, are only less important than those whose correlates are the vital forces recently derived from it.

§ 84. Regarded as an induction, the doctrine set forth in this chapter will most likely be met by a demurrer. Many who admit that among physical phenomena at least, the correlation of forces is now established, will probably say that inquiry has not yet gone far enough to enable us to predicate equivalence. And in respect of the forces classed as vital, mental, and social, the evidence assigned, however little to be explained away, they will consider by no means conclusive even of correlation, much less of equivalence.

To those who think thus, it must now however be pointed out, that the universal truth above illustrated under its various aspects, is a necessary corollary from the persistence of force. Setting out with the proposition that force can neither come into existence, nor cease to exist, the several foregoing general conclusions inevitably follow. Each manifestation of force can be interpreted only as the effect of some antecedent force: no matter whether it be an inorganic action, an animal movement, a thought, or a feeling. Either this must be conceded, or else it must be asserted that our successive states of consciousness are self-created. Either mental energies, as well as bodily ones, are quantitatively correlated to certain energies expended in their production, and to certain other energies which they initiate; or else nothing must become something and something must become nothing. The alternatives are, to deny the persistence of force, or to admit that every physical and psychial change is generated by certain antecedent forces, and that from given amounts of such forces neither more nor less of such physical and psychial changes can result. And since the persistence of force, being a datum of consciousness, cannot be denied, its unavoidable corollary must be accepted. This corollary cannot indeed be made more certain by accumulating illustrations. The truth as arrived at deductively, cannot be inductively confirmed. For every one of such facts as those above detailed, is established only through the indirect assumption of that persistence of force, from which it really follows as a direct consequence. The most exact proof of correlation and equivalence which it is possible to reach by experimental inquiry, is that based on measurement of the forces expended and the forces produced. But, as was shown in the last chapter, any such process of measurement implies the use of some unit of force which is assumed to remain constant; and for this assumption there can be no warrant but that it is a corollary from the persistence of force. How then can any reasoning based on this corollary, prove the equally direct corollary that when a given quantity of force ceases to exist under one form, an equal quantity must come into existence under some other form or forms? Clearly the *à priori* truth expressed in this last corollary, cannot be more firmly established by any *à posteriori* proofs which the first corollary helps us to.

"What then," it may be asked, "is the use of these investigations by which the correlation and equivalence of forces is sought to be established as an inductive truth? Surely it will not be alleged that they are useless. Yet if this correlation cannot be made more certain by them than it is already, does not their uselessness necessarily follow?" No. They are of value as disclosing the many particular implications which the general truth does not specify. They are of value as teaching us how much of one mode of force is the equivalent of so much of another mode. They are of value as determining under what conditions each metamorphosis occurs. And they are of value as leading us to inquire in what shape the remnant of force has escaped, when the apparent results are not equivalent to the cause.

CHAPTER X
THE DIRECTION OF MOTION

§ 85. The Absolute Cause of changes, inclusive of those constituting Evolution, is not less incomprehensible in respect of the unity or duality of its action, than in all other respects. We cannot decide between the alternative suppositions, that phenomena are due to the variously-conditioned workings of a single force, and that they are due to the conflict of two forces. Whether, as some contend, everything is explicable on the hypothesis of universal pressure, whence what we call tension results differentially from inequalities of pressure in opposite directions; or whether, as might be with equal propriety contended, things are to be explained on the hypothesis of universal tension, from which pressure is a differential result; or whether, as most physicists hold, pressure and tension everywhere co-exist; are questions which it is impossible to settle. Each of these three suppositions makes the facts comprehensible, only by postulating an inconceivability. To assume a universal pressure, confessedly requires us to assume an infinite plenum—an unlimited space full of something which is everywhere pressed by something beyond; and this assumption cannot be mentally realized. That universal tension is the immediate agency to which phenomena are due, is an idea open to a parallel and equally fatal objection. And however verbally intelligible may be the proposition that pressure and tension everywhere co-exist, yet we cannot truly represent to ourselves one ultimate unit of matter as drawing another while resisting it.

Nevertheless, this last belief is one which we are compelled to entertain. Matter cannot be conceived except as manifesting forces of attraction and repulsion. Body is distinguished in our consciousness from Space, by its opposition to our muscular energies; and this opposition we feel under the two-fold form of a cohesion that hinders our efforts to rend, and a resistance that hinders our efforts to compress. Without resistance there can be merely empty extension. Without cohesion there can be no resistance. Probably this conception of antagonistic forces, is originally derived from the antagonism of our flexor and extensor muscles. But be this as it may, we are obliged to think of all objects as made up of parts that attract and repel each other; since this is the form of our experience of all objects.

By a higher abstraction results the conception of attractive and repulsive forces pervading space. We cannot dissociate force from occupied extension, or occupied extension from force; because we have never an immediate consciousness of either in the absence of the other. Nevertheless, we have abundant proof that force is exercised through what appears to our senses a vacuity. Mentally to represent this exercise, we are hence obliged to fill the apparent vacuity with a species of matter—an etherial medium. The constitution we assign to this etherial medium, however, like the constitution we assign to solid substance, is necessarily an abstract of the impressions received from tangible bodies. The opposition to pressure which a tangible body offers to us, is not shown in one direction only, but in all directions; and so likewise is its tenacity. Suppose countless lines radiating from its centre on every side, and it resists along each of these lines and coheres along each of these lines. Hence the constitution of those ultimate units through the instrumentality of which phenomena are interpreted. Be they atoms of ponderable matter or molecules of ether, the properties we conceive them to possess are nothing else than these perceptible properties idealized. Centres of force attracting and repelling each other in all directions, are simply insensible portions of matter having the endowments common to sensible portions of matter—endowments of which we cannot by any mental effort divest them. In brief, they are the invariable elements of the conception of matter, abstracted from its variable elements—size, form, quality, &c. And so to interpret manifestations of force which cannot be tactually experienced, we use the terms of thought supplied by our tactual experiences; and this for the sufficient reason that we must use these or none.

After all that has been before shown, and after the hint given above, it needs scarcely be said that these universally co-existent forces of attraction and repulsion, must not be taken as realities, but as our symbols of the reality. They are the forms under which the workings of the Unknowable are cognizable by us—modes of the Unconditioned as presented under the conditions of our consciousness. But while knowing that the ideas thus generated in us are not absolutely true, we may unreservedly surrender ourselves to them as relatively true; and may proceed to evolve a series of deductions having a like relative truth.

§ 86. From universally co-existent forces of attraction and repulsion, there result certain laws of direction of all movement. Where attractive forces alone are concerned, or rather are alone appreciable, movement takes place in the direction of their resultant; which may, in a sense, be called the line of greatest traction. Where repulsive forces alone are concerned, or rather are alone appreciable, movement takes place along their resultant; which is usually known as the line of least resistance. And where both attractive and repulsive forces are concerned, or are appreciable, movement takes place along the resultant of all the tractions and resistances. Strictly speaking, this last is the sole law; since, by the hypothesis, both forces are everywhere in action. But very frequently the one kind of force is so immensely in excess that the effect of the other kind may be left out of consideration. Practically we may say that a body falling to the Earth, follows the line of greatest traction; since, though the resistance of the air must, if the body be irregular, cause some divergence from this line, (quite perceptible with feathers and leaves,) yet ordinarily the divergence is so slight that we may omit it. In the same manner, though the course taken by the steam from an exploding boiler, differs somewhat from that which it would take were gravitation out of the question; yet, as gravitation affects its course infinitesimally, we are justified in asserting that the escaping steam follows the line of least resistance. Motion then, we may say, always follows the line of greatest traction, or the line of least resistance, or the resultant of the two: bearing in mind that though the last is alone strictly true, the others are in many cases sufficiently near the truth for practical purposes.

Movement set up in any direction is itself a cause of further movement in that direction, since it is the embodiment of a surplus force in that direction. This holds equally with the transit of matter through space, the transit of matter through matter, and the transit through matter of any kind of vibration. In the case of matter moving through space, this principle is expressed in the law of inertia—a law on which the calculations of physical astronomy are wholly based. In the case of matter moving through matter, we trace the same truth under the familiar experience that any breach made by one solid through another, or any channel formed by a fluid through a solid, becomes a route along which, other things equal, subsequent movements of like nature take place. And in the case of motion passing through matter under the form of an impulse communicated from part to part, the facts of magnetization go to show that the establishment of undulations along certain lines, determines their continuance along those lines.

It further follows from the conditions, that the direction of movement can rarely if ever be perfectly straight. For matter in motion to pursue continuously the exact line in which it sets out, the forces of attraction and repulsion must be symmetrically disposed around its path; and the chances against this are infinitely great. The impossibility of making an absolutely true edge to a bar of metal—the fact that all which can be done by the best mechanical appliances, is to reduce the irregularities of such an edge to amounts that cannot be perceived without magnifiers—sufficiently exemplifies how, in consequence of the unsymmetrical distribution of forces around the line of movement, the movement is rendered more or less indirect. It may be well to add that in proportion as the forces at work are numerous and varied, the curve a moving body describes is necessarily complex: witness the contrast between the flight of an arrow and the gyrations of a stick tossed about by breakers.

We have now to trace these laws of direction of movement throughout the process of Evolution, under its various forms. We have to note how every change in the arrangement of parts, takes place along the line of greatest traction, of least resistance, or of their resultant; how the setting up of motion along a certain line, becomes a cause of its continuance along that line; how, nevertheless, change of relations to external forces, always renders this line indirect; and how the degree of its indirectness increases with every addition to the number of influences at work.

§ 87. If we assume the first stage in nebular condensation to be the precipitation into flocculi of denser matter previously diffused through a rarer medium, (a supposition both physically justified, and in harmony with certain astronomical observations,) we shall find that nebular motion is interpretable in pursuance of the above general laws. Each portion of such vapour-like matter must begin to move towards the common centre of gravity. The tractive forces which would of themselves carry it in a straight line to the centre of gravity, are opposed by the resistant forces of the medium through which it is drawn. The direction of movement must be the resultant of these—a resultant which, in consequence of the unsymmetrical form of the flocculus, must be a curve directed, not to the centre of gravity, but towards one side of it. And it may be readily shown that in an aggregation of such flocculi, severally thus moving, there must, by composition of forces, eventually result a rotation of the whole nebula in one direction.

Merely noting this hypothetical illustration for the purpose of showing how the law applies to the case of nebular evolution, supposing it to have taken place, let us pass to the phenomena of the Solar System as now exhibited. Here the general principles above set forth are every instant exemplified. Each planet and satellite has a momentum which would, if acting alone, carry it forward in the direction it is at any instant pursuing. This momentum hence acts as a resistance to motion in any other direction. Each planet and satellite, however, is drawn by a force which, if unopposed, would take it in a straight line towards its primary. And the resultant of these two forces is that curve which it describes—a curve manifestly consequent on the unsymmetrical distribution of the forces around its path. This path, when more closely examined, supplies us with further illustrations. For it is not an exact circle or ellipse; which it would be were the tangential and centripetal forces the only ones concerned. Adjacent members of the Solar System, ever varying in their relative positions, cause what we call perturbations; that is, slight divergences in various directions from that circle or ellipse which the two chief forces would produce. These perturbations severally show us in minor degrees, how the line of movement is the resultant of all the forces engaged; and how this line becomes more complicated in proportion as the forces are multiplied. If instead of the motions of the planets and satellites as wholes, we consider the motions of their parts, we meet with comparatively complex illustrations. Every portion of the Earth's substance in its daily rotation, describes a curve which is in the main a resultant of that resistance which checks its nearer approach to the centre of gravity, that momentum which would carry it off at a tangent, and those forces of gravitation and cohesion which keep it from being so carried off. If this axial motion be compounded with the orbital motion, the course of each part is seen to be a much more involved one. And we find it to have a still greater complication on taking into account that lunar attraction which mainly produces the tides and the precession of the equinoxes.

§ 88. We come next to terrestrial changes: present ones as observed, and past ones as inferred by geologists. Let us set out with the hourly-occurring alterations in the Earth's atmosphere; descend to the slower alterations in progress on its surface; and then to the still slower ones going on beneath.

Masses of air, absorbing heat from surfaces warmed by the sun, expand, and so lessen the weight of the atmospheric columns of which they are parts. Hence they offer to adjacent atmospheric columns, diminished lateral resistance; and these, moving in the directions of the diminished resistance, displace the expanded air; while this, pursuing an upward course, displays a motion along that line in which there is least pressure. When again, by the ascent of such heated masses from extended areas like the torrid zone, there is produced at the upper surface of the atmosphere, a protuberance beyond the limits of equilibrium—when the air forming this protuberance begins to overflow laterally towards the poles; it does so because, while the tractive force of the Earth is nearly the same, the lateral resistance is greatly diminished. And throughout the course of each current thus generated, as well as throughout the course of each counter-current flowing: into the vacuum that is left, the direction is always the resultant of the Earth's tractive force and the resistance offered by the surrounding masses of air: modified only by conflict with other currents similarly determined, and by collision with prominences on the Earth's crust. The movements of water, in both its gaseous and liquid states, furnish further examples. In conformity with the mechanical theory of heat, it may be shown that evaporation is the escape of particles of water in the direction of least resistance; and that as the resistance (which is due to the pressure of the water diffused in a gaseous state) diminishes, the evaporation increases. Conversely, that rushing together of particles called condensation, which takes place when any portion of atmospheric vapour has its temperature much lowered, may be interpreted as a diminution of the mutual pressure among the condensing particles, while the pressure of surrounding particles remains the same; and so is a motion taking place in the direction of lessened resistance. In the course followed by the resulting rain-drops, we have one of the simplest instances of the joint effect of the two antagonist forces. The Earth's attraction, and the resistance of atmospheric currents ever varying in direction and intensity, give as their resultants, lines which incline to the horizon in countless different degrees and undergo perpetual variations. More clearly still is the law exemplified by these same rain-drops when they reach the ground. In the course they take while trickling over its surface, in every rill, in every larger stream, and in every river, we see them descending as straight as the antagonism of surrounding objects permits. From moment to moment, the motion of water towards the Earth's centre is opposed by the solid matter around and under it; and from moment to moment its route is the resultant of the lines of greatest traction and least resistance. So far from a cascade furnishing, as it seems to do, an exception, it furnishes but another illustration. For though all solid obstacles to a vertical fall of the water are removed, yet the water's horizontal momentum is an obstacle; and the parabola in which the stream leaps from the projecting ledge, is generated by the combined gravitation and momentum. It may be well just to draw attention to the degree of complexity here produced in the line of movement by the variety of forces at work. In atmospheric currents, and still more clearly in water-courses (to which might be added ocean-streams), the route followed is too complex to be defined, save as a curve of three dimensions with an ever varying equation.

The Earth's solid crust undergoes changes that supply another group of illustrations. The denudation of lands and the depositing of the removed sediment in new strata at the bottoms of seas and lakes, is a process throughout which motion is obviously determined in the same way as is that of the water effecting the transport. Again, though we have no direct inductive proof that the forces classed as igneous, expend themselves along lines of least resistance; yet what little we know of them is in harmony with the belief that they do so. Earthquakes continually revisit the same localities, and special tracts undergo for long periods together successive elevations or subsidences,—facts which imply that already-fractured portions of the Earth's crust are those most prone to yield under the pressure caused by further contractions. The distribution of volcanoes along certain lines, as well as the frequent recurrence of eruptions from the same vents, are facts of like meaning.

§ 89. That organic growth takes place in the direction of least resistance, is a proposition that has been set forth and illustrated by Mr. James Hinton, in the *Medico-Chirurgical Review* for October, 1858. After detailing a few of the early observations which led him to this generalization, he formulates it thus: —

"Organic form is the result of motion."

"Motion takes the direction of least resistance."

"Therefore organic form is the result of motion in the direction of least resistance."

After an elucidation and defence of this position, Mr. Hinton proceeds to interpret, in conformity with it, sundry phenomena of development. Speaking of plants he says: —

"The formation of the root furnishes a beautiful illustration of the law of least resistance, for it grows by insinuating itself, cell by cell, through the interstices of the soil; it is by such minute additions that it increases, winding and twisting whithersoever the obstacles it meets in its path determine, and growing there most, where the nutritive materials are added to it most abundantly. As we look on the roots of a mighty tree, it appears to us as if they had forced themselves with giant violence into the solid earth. But it is not so; they were led on gently, cell added to cell, softly as the dews descended, and the loosened earth made way. Once formed, indeed, they expand with an enormous power, but the spongy condition of the growing radicles utterly forbids the supposition that they are forced into the earth. Is it not probable, indeed, that the enlargement of the roots already formed may crack the surrounding soil, and help to make the interstices into which the new rootlets grow?"

"Throughout almost the whole of organic nature the spiral form is more or less distinctly marked. Now, motion under resistance takes a spiral direction, as may be seen by the motion of a body rising or falling through water. A bubble rising rapidly in water describes a spiral closely resembling a corkscrew, and a body of moderate specific gravity dropped into water may be seen to fall in a curved direction, the spiral tendency of which may be distinctly observed.In this prevailing spiral form of organic bodies, therefore, it appears to me, that there is presented a strong *prima facie* case for the view I have maintained.The spiral form of the branches of many trees is very apparent, and the universally spiral arrangement of the leaves around the stem of plants needs only to be referred to. * * * The heart commences as a spiral turn, and in its perfect form a manifest spiral may be traced through the left ventricle, right ventricle, right auricle, left auricle and appendix. And what is the spiral turn in which the heart commences but a necessary result of the lengthening, under a limit, of the cellular mass of which it then consists?"

"Every one must have noticed the peculiar curling up of the young leaves of the common fern. The appearance is as if the leaf were rolled up, but in truth this form is merely a phenomenon of growth. The curvature results from the increase of the leaf, it is only another form of the wrinkling up, or turning at right angles by extension under limit."

"The rolling up or imbrication of the petals in many flower-buds is a similar thing; at an early period the small petals may be seen lying side by side, afterwards growing within the capsule, they become folded round one another."

"If a flower-bud be opened at a sufficiently early period, the stamens will be found as if moulded in the cavity between the pistil and the corolla, which cavity the antlers exactly fill; the stalks lengthen at an after period. I have noticed also in a few instances, that in those flowers in which the petals are imbricated, or twisted together, the pistil is tapering as growing up between the petals; in some flowers which have the petals so arranged in the bud as to form a dome (as the hawthorn; e. g.), the pistil is flattened at the apex, and in the bud occupies a space precisely limited by the stamens below, and the enclosing petals above and at the sides. I have not, however, satisfied myself that this holds good in all cases."

Without endorsing all Mr. Hinton's illustrations, to some of which exception might be taken, his conclusion may be accepted as a large instalment of the truth. It is, however, to be remarked, that in the case of organic growth, as in all other cases, the line of movement is in strictness the resultant of tractive and resistant forces; and that the tractive forces here form so considerable an element that the formula is scarcely complete without them. The shapes of plants are manifestly modified by gravitation: the direction of each branch is not what it would have been were the tractive force of the Earth absent; and every flower and leaf is somewhat altered in the course of development by the weight of its parts. Though in animals such effects are less conspicuous, yet the instances in which flexible organs have their directions in great measure determined by gravity, justify the assertion that throughout the whole organism the forms of parts must be affected by this force.

The organic movements which constitute growth, are not, however, the only organic movements to be interpreted. There are also those which constitute function. And throughout these the same general principles are discernible. That the vessels along which blood, lymph, bile, and all the secretions, find their ways, are channels of least resistance, is a fact almost too conspicuous to be named as an illustration. Less conspicuous, however, is the truth, that the currents setting along these vessels are affected by the tractive force of the Earth: witness varicose veins; witness the relief to an inflamed part obtained by raising it; witness the congestion of head and face produced by stooping. And in the fact that dropsy in the legs gets greater by day and decreases at night, while, conversely, that œdematous fullness under the eyes common in debility, grows worse during the hours of reclining and decreases after getting up, shows us how the transudation of fluid through the walls of the capillaries, varies according as change of position changes the effect of gravity in different parts of the body.

It may be well in passing just to note the bearing of the principle on the development of species. From a dynamic point of view, "natural selection" is the evolution of Life along lines of least resistance. The multiplication of any kind of plant or animal in localities that are favourable to it, is a growth where the antagonistic forces are less than elsewhere. And the preservation of varieties that succeed better than their allies in coping with surrounding conditions, is the continuance of vital movement in those directions where the obstacles to it are most eluded.

§ 90. Throughout the phenomena of mind the law enunciated is not so readily established. In a large part of them, as those of thought and emotion, there is no perceptible movement. Even in sensation and volition, which show us in one part of the body an effect produced by a force applied to another part, the intermediate movement is inferential rather than visible. Such indeed are the difficulties that it is not possible here to do more than briefly indicate the proofs which might be given did space permit.

Supposing the various forces throughout an organism to be previously in equilibrium, then any part which becomes the seat of a further force, added or liberated, must be one from which the force, being resisted by smaller forces around, will initiate motion towards some other part of the organism. If elsewhere in the organism there is a point at which force is being expended, and which so is becoming minus a force which it before had, instead of plus a force which it before had not, and thus is made a point at which the re-action against surrounding forces is diminished; then, manifestly, a motion taking place between the first and the last of these points is a motion along the line of least resistance. Now a sensation implies a force added to, or evolved in, that part of the organism which is its seat; while a mechanical movement implies an expenditure or loss of force in that part of the organism which is its seat. Hence if, as we find to be the fact, motion is habitually propagated from those parts of an organism to which the external world adds forces in the shape of nervous impressions, to those parts of an organism which react on the external world through muscular contractions, it is simply a fulfilment of the law above enunciated. From this general conclusion we may pass to a more special one. When there is anything in the circumstances of an animal's life, involving that a sensation in one particular place is habitually followed by a contraction in another particular place—when there is thus a frequently-repeated motion through the organism between these places; what must be the result as respects the line along which the motions take place? Restoration of equilibrium between the points at which the forces have been increased and decreased, must take place through some channel. If this channel is affected by the discharge—if the obstructive action of the tissues traversed, involves any reaction upon them, deducting from their obstructive power; then a subsequent motion between these two points will meet with less resistance along this channel than the previous motion met with; and will consequently take this channel still more decidedly. If so, every repetition will still further diminish the resistance offered by this route; and hence will gradually be formed between the two a permanent line of communication, differing greatly from the surrounding tissue in respect of the ease with which force traverses it. We see, therefore, that if between a particular impression and a particular motion associated with it, there is established a connexion producing what is called reflex action, the law that motion follows the line of least resistance, and that, if the conditions remain constant, resistance in any direction is diminished by motion occurring in that direction, supplies an explanation. Without further details it will be manifest that a like interpretation may be given to the succession of all other nervous changes. If in the surrounding world there are objects, attributes, or actions, that usually occur together, the effects severally produced by them in the organism will become so connected by those repetitions which we call experience, that they also will occur together. In proportion to the frequency with which any external connexion of phenomena is experienced, will be the strength of the answering internal connexion of nervous states. Thus there will arise all degrees of cohesion among nervous states, as there are all degrees of commonness among the surrounding co-existences and sequences that generate them: whence must result a general correspondence between associated ideas and associated actions in the environment.[13]

The relation between emotions and actions may be similarly construed. As a first illustration let us observe what happens with emotions that are undirected by volitions. These, like feelings in general, expend themselves in generating organic changes, and chiefly in muscular contractions. As was pointed out in the last chapter, there result movements of the involuntary and voluntary muscles, that are great in proportion as the emotions are strong. It remains here to be pointed out, however, that the order in which these muscles are affected is explicable only on the principle above set forth. Thus, a pleasurable or painful state of mind of but slight intensity, does little more than increase the pulsations of the heart. Why? For the reason that the relation between nervous excitement and vascular contraction, being common to every genus and species of feeling, is the one of most frequent repetition; that hence the nervous connexion is, in the way above shown, the one which offers the least resistance to a discharge; and is therefore the one along which a feeble force produces motion. A sentiment or passion that is somewhat stronger, affects not only the heart but the muscles of the face, and especially those around the mouth. Here the like explanation applies; since these muscles, being both comparatively small, and, for purposes of speech, perpetually used, offer less resistance than other voluntary muscles to the nerve-motor force. By a further increase of emotion the respiratory and vocal muscles become perceptibly excited. Finally, under strong passion, the muscles in general of the trunk and limbs are violently contracted. Without saying that the facts can be thus interpreted in all their details (a task requiring data impossible to obtain) it may be safely said that the order of excitation is from muscles that are small and frequently acted on, to those which are larger and less frequently acted on. The single instance of laughter, which is an undirected discharge of feeling that affects first the muscles round the mouth, then those of the vocal and respiratory apparatus, then those of the limbs, and then those of the spine;[14] suffices to show that when no special route is opened for it, a force evolved in the nervous centres produces motion along channels which offer the least resistance, and if it is too great to escape by these, produces motion along channels offering successively greater resistance.

Probably it will be thought impossible to extend this reasoning so as to include volitions. Yet we are not without evidence that the transition from special desires to special muscular acts, conforms to the same principle. It may be shown that the mental antecedents of a voluntary movement, are antecedents which temporarily make the line along which this movement takes place, the line of least resistance. For a volition, suggested as it necessarily is by some previous thought connected with it by associations that determine the transition, is itself a representation of the movements that are willed, and of their sequences. But to represent in consciousness certain of our own movements, is partially to arouse the sensations accompanying such movements, inclusive of those of muscular tension—is partially to excite the appropriate motor-nerves and all the other nerves implicated. That is to say, the volition is itself an incipient discharge along a line which previous experiences have rendered a line of least resistance. And the passing of volition into action is simply a completion of the discharge.

One corollary from this must be noted before proceeding; namely, that the particular set of muscular movements by which any object of desire is reached, are movements implying the smallest total of forces to be overcome. As each feeling generates motion along the line of least resistance, it is tolerably clear that a group of feelings, constituting a more or less complex desire, will generate motion along a series of lines of least resistance. That is to say, the desired end will be achieved with the smallest expenditure of effort. Should it be objected that through want of knowledge or want of skill, a man often pursues the more laborious of two courses, and so overcomes a larger total of opposing forces than was necessary; the reply is, that relatively to his mental state the course he takes is that which presents the fewest difficulties. Though there is another which in the abstract is easier, yet his ignorance of it, or inability to adopt it, is, physically considered, the existence of an insuperable obstacle to the discharge of his energies in that direction. Experience obtained by himself, or communicated by others, has not established in him such channels of nervous communication as are required to make this better course the course of least resistance to him.

§ 91. As in individual animals, inclusive of man, motion follows lines of least resistance, it is to be inferred that among aggregations of men, the like will hold good. The changes in a society, being due to the joint actions of its members, the courses of such changes will be determined as are those of all other changes wrought by composition of forces.

Thus when we contemplate a society as an organism, and observe the direction of its growth, we find this direction to be that in which the average of opposing forces is the least. Its units have energies to be expended in self-maintenance and reproduction. These energies are met by various environing energies that are antagonistic to them—those of geological origin, those of climate, of wild animals, of other human races with whom they are at enmity or in competition. And the tracts the society spreads over, are those in which there is the smallest total antagonism. Or, reducing the matter to its ultimate terms, we may say that these social units have jointly and severally to preserve themselves and their offspring from those inorganic and organic forces which are ever tending to destroy them (either indirectly by oxidation and by undue abstraction of heat, or directly by bodily mutilation); that these forces are either counteracted by others which are available in the shape of food, clothing, habitations, and appliances of defence, or are, as far as may be, eluded; and that population spreads in whichever directions there is the readiest escape from these forces, or the least exertion in obtaining the materials for resisting them, or both. For these reasons it happens that fertile valleys where water and vegetal produce abound, are early peopled. Sea-shores, too, supplying a large amount of easily-gathered food, are lines along which mankind have commonly spread. The general fact that, so far as we can judge from the traces left by them, large societies first appeared in those tropical regions where the fruits of the earth are obtainable with comparatively little exertion, and where the cost of maintaining bodily heat is but slight, is a fact of like meaning. And to these instances may be added the allied one daily furnished by emigration; which we see going on towards countries presenting the fewest obstacles to the self-preservation of individuals, and therefore to national growth. Similarly with that resistance to the movements of a society which neighbouring societies offer. Each of the tribes or nations inhabiting any region, increases in numbers until it outgrows its means of subsistence. In each there is thus a force ever pressing outwards on to adjacent areas—a force antagonized by like forces in the tribes or nations occupying those areas. And the ever-recurring wars that result—the conquests of weaker tribes or nations, and the over-running of their territories by the victors, are instances of social movements taking place in the directions of least resistance. Nor do the conquered peoples, when they escape extermination or enslavement, fail to show us movements that are similarly determined. For migrating as they do to less fertile regions—taking refuge in deserts or among mountains—moving in a direction where the resistance to social growth is comparatively great; they still do this only under an excess of pressure in all other directions: the physical obstacles to self-preservation they encounter, being really less than the obstacles offered by the enemies from whom they fly.

Internal social movements may also be thus interpreted. Localities naturally fitted for producing particular commodities—that is, localities in which such commodities are got at the least cost of force—that is, localities in which the desires for these commodities meet with the least resistance; become localities especially devoted to the obtainment of these commodities. Where soil and climate render wheat a profitable crop, or a crop from which the greatest amount of life-sustaining power is gained by a given quantity of effort, the growth of wheat becomes the dominant industry. Where wheat cannot be economically produced, oats, or rye, or maize, or rice, or potatoes, is the agricultural staple. Along sea-shores men support themselves with least effort by catching fish; and hence choose fishing as an occupation. And in places that are rich in coal or metallic ores, the population, finding that labour devoted to the raising of these materials brings a larger return of food and clothing than when otherwise directed, becomes a population of miners. This last instance introduces us to the phenomena of exchange; which equally illustrate the general law. For the practice of barter begins as soon as it facilitates the fulfilment of men's desires, by diminishing the exertion needed to reach the objects of those desires. When instead of growing his own corn, weaving his own cloth, sewing his own shoes, each man began to confine himself to farming, or weaving, or shoemaking; it was because each found it more laborious to make everything he wanted, than to make a great quantity of one thing and barter the surplus for the rest: by exchange, each procured the necessaries of life without encountering so much resistance. Moreover, in deciding what commodity to produce, each citizen was, as he is at the present day, guided in the same manner. For besides those local conditions which determine whole sections of a society towards the industries easiest for them, there are also individual conditions and individual aptitudes which to each citizen render certain occupations preferable; and in choosing those forms of activity which their special circumstances and faculties dictate, these social units are severally moving towards the objects of their desires in the directions which present to them the fewest obstacles. The process of transfer which commerce pre-supposes, supplies another series of examples. So long as the forces to be overcome in procuring any necessary of life in the district where it is consumed, are less than the forces to be overcome in procuring it from an adjacent district, exchange does not take place. But when the adjacent district produces it with an economy that is not out-balanced by cost of transit—when the distance is so small and the route so easy that the labour of conveyance plus the labour of production is less than the labour of production in the consuming district, transfer commences. Movement in the direction of least resistance is also seen in the establishment of the channels along which intercourse takes place. At the outset, when goods are carried on the backs of men and horses, the paths chosen are those which combine shortness with levelness and freedom from obstacles—those which are achieved with the smallest exertion. And in the subsequent formation of each highway, the course taken is that which deviates horizontally from a straight line so far only as is needful to avoid vertical deviations entailing greater labour in draught. The smallest total of obstructive forces determines the route, even in seemingly exceptional cases; as where a detour is made to avoid the opposition of a landowner. All subsequent improvements, ending in macadamized roads, canals, and railways, which reduce the antagonism of friction and gravity to a minimum, exemplify the same truth. After there comes to be a choice of roads between one point and another, we still see that the road chosen is that along which the cost of transit is the least: cost being the measure of resistance. Even where, time being a consideration, the more expensive route is followed, it is so because the loss of time involves loss of force. When, division of labour having been carried to a considerable extent and means of communication made easy, there arises a marked localization of industries, the relative growths of the populations devoted to them may be interpreted on the same principle. The influx of people to each industrial centre, as well as the rate of multiplication of those already inhabiting it, is determined by the payment for labour; that is—by the quantity of commodities which a given amount of effort will obtain. To say that artisans flock to places where, in consequence of facilities for production, an extra proportion of produce can be given in the shape of wages; is to say that they flock to places where there are the smallest obstacles to the support of themselves and families. Hence, the rapid increase of number which occurs in such places, is really a social growth at points where the opposing forces are the least.

Nor is the law less clearly to be traced in those functional changes daily going on. The flow of capital into businesses yielding the largest returns; the buying in the cheapest market and selling in the dearest; the introduction of more economical modes of manufacture; the development of better agencies for distribution; and all those variations in the currents of trade that are noted in our newspapers and telegrams from hour to hour; exhibit movement taking place in directions where it is met by the smallest total of opposing forces. For if we analyze each of these changes—if instead of interest on capital we read surplus of products which remains after maintenance of labourers; if we so interpret large interest or large surplus to imply labour expended with the greatest results; and if labour expended with the greatest results means muscular action so directed as to evade obstacles as far as possible; we see that all these commercial phenomena are complicated motions set up along lines of least resistance.

Objections of two opposite kinds will perhaps be made to these sociological applications of the law. By some it may be said that the term force as here used, is used metaphorically—that to speak of men as *impelled* in certain directions by certain desires, is a figure of speech and not the statement of a physical fact. The reply is, that the foregoing illustrations are to be interpreted literally, and that the processes described *are* physical ones. The pressure of hunger is an actual force—a sensation implying some state of nervous tension; and the muscular action which the sensation prompts is really a discharge of it in the shape of bodily motion—a discharge which, on analyzing the mental acts involved, will be found to follow lines of least resistance. Hence the motions of a society whose members are impelled by this or any other desire, are actually, and not metaphorically, to be understood in the manner shown. An opposite objection may possibly be, that the several illustrations given are elaborated truisms; and that the law of direction of motion being once recognized, the fact that social movements, in common with all others, must conform to it, follows inevitably. To this it may be rejoined, that a mere abstract assertion that social movements must do this, would carry no conviction to the majority; and that it is needful to show *how* they do it. For social evolution to be interpreted after the method proposed, it is requisite that such generalisations as those of political economy shall be reduced to equivalent propositions expressed in terms of force and motion.

Social movements of these various orders severally conform to the two derivative principles named at the outset. In the first place we may observe how, once set up in given directions, such movements, like all others, tend to continue in these directions. A commercial mania or panic, a current of commodities, a social custom, a political agitation, or a popular delusion, maintains its course for a long time after its original source has ceased; and requires antagonistic forces to arrest it. In the second place it is to be noted that in proportion to the complexity of social forces is the tortuousness of social movements. The involved series of muscular contractions gone through by the artizan, that he may get the wherewithal to buy a loaf lying at the baker's next door, show us how extreme becomes the indirectness of motion when the agencies at work become very numerous—a truth still better illustrated by the more public social actions; as those which end in bringing a successful man of business, towards the close of his life, into parliament.

§ 92. And now of the general truth set forth in this chapter, as of that dealt with in the last, let us ask—what is our ultimate evidence? Must we accept it simply as an empirical generalization? or is it to be established as a corollary from a still deeper truth? The reader will anticipate the answer. We shall find it deducible from that datum of consciousness which underlies all science.

Suppose several tractive forces, variously directed, to be acting on a given body. By what is known among mathematicians as the composition of forces, there may be found for any two of these, a single force of such amount and direction as to produce on the body an exactly equal effect. If in the direction of each of them there be drawn a straight line, and if the lengths of these two straight lines be made proportionate to the amounts of the forces; and if from the end of each line there be drawn a line parallel to the other, so as to complete a parallelogram; then the diagonal of this parallelogram represents the amount and direction of a force that is equivalent to the two.

Such a resultant force, as it is called, may be found for any pair of forces throughout the group. Similarly, for any pair of such resultants a single resultant may be found. And by repeating this course, all of them may be reduced to two. If these two are equal and opposite—that is, if there is no line of greatest traction, motion does not take place. If they are opposite but not equal, motion takes place in the direction of the greater. And if they are neither equal nor opposite, motion takes place in the direction of their resultant. For in either of these cases there is an unantagonized force in one direction. And this residuary force that is not neutralized by an opposing one, must move the body in the direction in which it is acting. To assert the contrary is to assert that a force can be expended without effect—without generating an equivalent force; and by so implying that force can cease to exist, this involves a denial of the persistence of force. It needs scarcely be added that if in place of tractions we take resistances, the argument equally holds; and that it holds also where both tractions and resistances are concerned. Thus the law that motion follows the line of greatest traction, or the line of least resistance, or the resultant of the two, is a necessary deduction from that primordial truth which transcends proof.

Reduce the proposition to its simplest form, and it becomes still more obviously consequent on the persistence of force. Suppose two weights suspended over a pulley or from the ends of an equal-armed lever; or better still—suppose two men pulling against each other. In such cases we say that the heavier weight will descend, and that the stronger man will draw the weaker towards him. But now, if we are asked how we know which is the heavier weight or the stronger man; we can only reply that it is the one producing motion in the direction of its pull. Our only evidence of excess of force is the movement it produces. But if of two opposing tractions we can know one as greater than the other only by the motion it generates in its own direction, then the assertion that motion occurs in the direction of greatest traction is a truism. When, going a step further back, we seek a warrant for the assumption that of the two conflicting forces, that is the greater which produces motion in its own direction, we find no other than the consciousness that such part of the greater force as is unneutralized by the lesser, must produce its effect—the consciousness that this residuary force cannot disappear, but must manifest itself in some equivalent change—the consciousness that force is persistent. Here too, as before, it may be remarked that no amount of varied illustrations, like those of which this chapter mainly consists, can give greater certainty to the conclusion thus immediately drawn from the ultimate datum of consciousness. For in all cases, as in the simple ones just given, we can identify the greatest force only by the resulting motion. It is impossible for us ever to get evidence of the occurrence of motion in any other direction than that of the greatest force; since our measure of relative greatness among forces is their relative power of generating motion. And clearly, while the comparative greatness of forces is thus determined, no multiplication of instances can add certainty to a law of direction of movement which follows immediately from the persistence of force.

From this same primordial truth, too, may be deduced the principle that motion once set up along any line, becomes itself a cause of subsequent motion along that line. The mechanical axiom that, if left to itself, matter moving in any direction will continue in that direction with undiminished velocity, is but an indirect assertion of the persistence of force; since it is an assertion that the force manifested in the transfer of a body along a certain length of a certain line in a certain time, cannot disappear without producing some equal manifestation—a manifestation which, in the absence of conflicting forces, must be a further transfer in the same direction at the same velocity. In the case of matter traversing matter the like inference is necessitated. Here indeed the actions are much more complicated. A liquid that follows a certain channel through or over a solid, as water along the Earth's surface, loses part of its motion in the shape of heat, through friction and collision with the matters forming its bed. A further amount of its motion may be absorbed in overcoming forces which it liberates; as when it loosens a mass which falls into, and blocks up, its channel. But after these deductions by transformation into other modes of force, any further deduction from the motion of the water is at the expense of a reaction on the channel, which by so much diminishes its obstructive power: such reaction being shown in the motion acquired by the detached portions which are carried away. The cutting out of river-courses is a perpetual illustration of this truth. Still more involved is the case of motion passing through matter by impulse from part to part; as a nervous discharge through animal tissue. Some chemical change may be wrought along the route traversed, which may render it less fit than before for conveying a current. Or the motion may itself be in part metamorphosed into some obstructive form of force; as in metals, the conducting power of which is, for the time, decreased by the heat which the passage of electricity itself generates. The real question is, however, what structural modification, if any, is produced throughout the matter traversed, apart from *incidental* disturbing forces—apart from everything but the *necessary* resistance of the matter: that, namely, which results from the inertia of its units. If we confine our attention to that part of the motion which, escaping transformation, continues its course, then it is a corollary from the persistence of force that as much of this remaining motion as is taken up in changing the positions of the units, must leave these by so much less able to obstruct subsequent motion in the same direction.

Thus in all the changes heretofore and at present displayed by the Solar System; in all those that have gone on and are still going on in the Earth's crust; in all processes of organic development and function; in all mental actions and the effects they work on the body; and in all modifications of structure and activity in societies; the implied movements are of necessity determined in the manner above set forth. Every alteration in the arrangement of parts, constituting Evolution under each of its phases, must conform to this universal principle. Wherever we see motion, its direction must be that of the greatest force. And wherever we see the greatest force to be acting in a given direction, in that direction motion must ensue.

> 13. This paragraph is a re-statement, somewhat amplified, of an idea set forth in the *Medico-Chirurgical Review* for January, 1859 (pp. 189 and 190); and contains the germ of the intended fifth part of the *Principles of Psychology*, which was withheld for the reasons given in the preface to that work.

> 14. For details see a paper on "The Physiology of Laughter," published in *Macmillan's Magazine* for March 1860.

CHAPTER XI
THE RHYTHM OF MOTION

§ 93. When the pennant of a vessel lying becalmed first shows the coming breeze, it does so by gentle undulations that travel from its fixed to its free end. Presently the sails begin to flap; and their blows against the mast increase in rapidity as the breeze rises. Even when, being fully bellied out, they are in great part steadied by the strain of the yards and cordage, their free edges tremble with each stronger gust. And should there come a gale, the jar that is felt on laying hold of the shrouds shows that the rigging vibrates; while the rush and whistle of the wind prove that in it, also, rapid undulations are generated. Ashore the conflict between the current of air and the things it meets results in a like rhythmical action. The leaves all shiver in the blast; each branch oscillates; and every exposed tree sways to and fro. The blades of grass and dried bents in the meadows, and still better the stalks in the neighbouring corn-fields, exhibit the same rising and falling movement. Nor do the more stable objects fail to do the like, though in a less manifest fashion; as witness the shudder that may be felt throughout a house during the paroxysms of a violent storm. Streams of water produce in opposing objects the same general effects as do streams of air. Submerged weeds growing in the middle of a brook, undulate from end to end. Branches brought down by the last flood, and left entangled at the bottom where the current is rapid, are thrown into a state of up and down movement that is slow or quick in proportion as they are large or small; and where, as in great rivers like the Mississippi, whole trees are thus held, the name "sawyers," by which they are locally known, sufficiently describes the rhythm produced in them. Note again the effect of the antagonism between the current and its channel. In shallow places, where the action of the bottom on the water flowing over it is visible, we see a ripple produced—a series of undulations. And if we study the action and re-action going on between the moving fluid and its banks, we still find the principle illustrated, though in a different way. For in every rivulet, as in the mapped-out course of every great river, the bends of the stream from side to side throughout its tortuous course constitute a lateral undulation—an undulation so inevitable that even an artificially straightened channel is eventually changed into a serpentine one. Analogous phenomena may be observed where the water is stationary and the solid matter moving. A stick drawn laterally through the water with much force, proves by the throb which it communicates to the hand that it is in a state of vibration. Even where the moving body is massive, it only requires that great force should be applied to get a sensible effect of like kind: instance the screw of a screw-steamer, which instead of a smooth rotation falls into a rapid rhythm that sends a tremor through the whole vessel. The sound which results when a bow is drawn over a violin-string, shows us vibrations produced by the movement of a solid over a solid. In lathes and planing machines, the attempt to take off a thick shaving causes a violent jar of the whole apparatus, and the production of a series of waves on the iron or wood that is cut. Every boy in scraping his slate-pencil finds it scarcely possible to help making a ridged surface. If you roll a ball along the ground or over the ice, there is always more or less up and down movement—a movement that is visible while the velocity is considerable, but becomes too small and rapid to be seen by the unaided eye as the velocity diminishes. However smooth the rails, and however perfectly built the carriages, a railway-train inevitably gets into oscillations, both lateral and vertical. Even where moving matter is suddenly arrested by collision, the law is still illustrated; for both the body striking and the body struck are made to tremble; and trembling is rhythmical movement. Little as we habitually observe it, it is yet certain that the impulses our actions impress from moment to moment on surrounding objects, are propagated through them in vibrations. It needs but to look through a telescope of high power, to be convinced that each pulsation of the heart gives a jar to the whole room. If we pass to motions of another order—those namely which take place in the etherial medium—we still find the same thing. Every fresh discovery confirms the hypothesis that light consists of undulations. The rays of heat, too, are now found to have a like fundamental nature; their undulations differing from those of light only in their comparative length. Nor do the movements of electricity fail to furnish us with an illustration; though one of a different order. The northern aurora may often be observed to pulsate with waves of greater brightness; and the electric discharge through a vacuum shows us by its stratified appearance that the current is not uniform, but comes in gushes of greater and lesser intensity. Should it be said that at any rate there are some motions, as those of projectiles, which are not rhythmical, the reply is, that the exception is apparent only; and that these motions would be rhythmical if they were not interrupted. It is common to assert that the trajectory of a cannon ball is a parabola; and it is true that (omitting atmospheric resistance) the curve described differs so slightly from a parabola that it may practically be regarded as one. But, strictly speaking, it is a portion of an extremely eccentric ellipse, having the Earth's centre of gravity for its remoter focus; and but for its arrest by the substance of the Earth, the cannon ball would travel round that focus and return to the point whence it started; again to repeat this slow rhythm. Indeed, while seeming at first sight to do the reverse, the discharge of a cannon furnishes one of the best illustrations of the principle enunciated. The explosion produces violent undulations in the surrounding air. The whizz of the shot, as it flies towards its mark, is due to another series of atmospheric undulations. And the movement to and from the Earth's centre, which the cannon ball is beginning to perform, being checked by solid matter, is transformed into a rhythm of another order; namely, the vibration which the blow sends through neighbouring bodies.[15]

Rhythm is very generally not simple but compound. There are usually at work various forces, causing undulations differing in rapidity; and hence it continually happens that besides the primary rhythms there are secondary rhythms, produced by the periodic coincidence and antagonism of the primary ones. Double, triple, and even quadruple rhythms, are thus generated. One of the simplest instances is afforded by what in acoustics are known as "beats:" recurring intervals of sound and silence which are perceived when two notes of nearly the same pitch are struck together; and which are due to the alternate correspondence and antagonism of the atmospheric waves. In like manner the various phenomena due to what is called interference of light, severally result from the periodic agreement and disagreement of etherial undulations—undulations which, by alternately intensifying and neutralizing each other, produce intervals of increased and diminished light. On the sea-shore may be noted sundry instances of compound rhythm. We have that of the tides, in which the daily rise and fall undergoes a fortnightly increase and decrease, due to the alternate coincidence and antagonism of the solar and lunar attractions. We have again that which is perpetually furnished by the surface of the sea: every large wave bearing smaller ones on its sides, and these still smaller ones; with the result that each flake of foam, along with the portion of water bearing it, undergoes minor ascents and descents of several orders while it is being raised and lowered by the greater billows. A quite different and very interesting example of compound rhythm, occurs in the little rills which, at low tide, run over the sand out of the shingle banks above. Where the channel of one of these is narrow, and the stream runs strongly, the sand at the bottom is raised into a series of ridges corresponding to the ripple of the water. On watching for a short time, it will be seen that these ridges are being raised higher and the ripple growing stronger; until at length, the action becoming violent, the whole series of ridges is suddenly swept away, the stream runs smoothly, and the process commences afresh. Instances of still more complex rhythms might be added; but they will come more appropriately in connexion with the several forms of Evolution, hereafter to be dealt with.

From the ensemble of the facts as above set forth, it will be seen that rhythm results wherever there is a conflict of forces not in equilibrium. If the antagonist forces at any point are balanced, there is rest; and in the absence of motion there can of course be no rhythm. But if instead of a balance there is an excess of force in one direction—if, as necessarily follows, motion is set up in that direction; then for that motion to continue uniformly in that direction, it is requisite that the moving matter should, notwithstanding its unceasing change of place, present unchanging relations to the sources of force by which its motion is produced and opposed. This however is impossible. Every further transfer through space must alter the ratio between the forces concerned—must increase or decrease the predominance of one force over the other—must prevent uniformity of movement. And if the movement cannot be uniform, then, in the absence of acceleration or retardation continued through infinite time and space, (results which cannot be conceived) the only alternative is rhythm.

A secondary conclusion must not be omitted. In the last chapter we saw that motion is never absolutely rectilinear; and here it remains to be added that, as a consequence, rhythm is necessarily incomplete. A truly rectilinear rhythm can arise only when the opposing forces are in exactly the same line; and the probabilities against this are infinitely great. To generate a perfectly circular rhythm, the two forces concerned must be exactly at right angles to each other, and must have exactly a certain ratio; and against this the probabilities are likewise infinitely great. All other proportions and directions of the two forces will produce an ellipse of greater or less eccentricity. And when, as indeed always happens, above two forces are engaged, the curve described must be more complex; and cannot exactly repeat itself. So that in fact throughout nature, this action and re-action of forces never brings about a complete return to a previous state. Where the movement is very involved, and especially where it is that of some aggregate whose units are partially independent, anything like a regular curve is no longer traceable; we see nothing more than a general oscillation. And on the completion of any periodic movement, the degree in which the state arrived at differs from the state departed from, is usually marked in proportion as the influences at work are numerous.

§ 94. That spiral arrangement so general among the more diffused nebulæ—an arrangement which must be assumed by matter moving towards a centre of gravity through a resisting medium—shows us the progressive establishment of revolution, and therefore of rhythm; in those remote spaces which the nebulæ occupy. Double stars, moving round common centres of gravity in periods some of which are now ascertained, exhibit settled rhythmical actions in distant parts of our sidereal system. And another fact which, though of a different order, has a like general significance, is furnished by variable stars—stars which alternately brighten and fade.

The periodicities of the planets, satellites, and comets, are so familiar that it would be inexcusable to name them, were it not needful here to point out that they are so many grand illustrations of this general law of movement. But besides the revolutions of these bodies in their orbits (all more or less excentric) and their rotations on their axes, the Solar System presents us with various rhythms of a less manifest and more complex kind. In each planet and satellite there is the revolution of the nodes—a slow change in the position of the orbit-plane, which after completing itself commences afresh. There is the gradual alteration in the length of the axis major of the orbit; and also of its excentricity: both of which are rhythmical alike in the sense that they alternate between maxima and minima, and in the sense that the progress from one extreme to the other is not uniform, but is made with fluctuating velocity. Then, too, there is the revolution of the line of apsides, which in course of time moves round the heavens—not regularly, but through complex oscillations. And further we have variations in the directions of the planetary axes—that known as nutation, and that larger gyration which, in the case of the Earth, causes the precession of the equinoxes. These rhythms, already more or less compound, are compounded with each other. Such an instance as the secular acceleration and retardation of the moon, consequent on the varying excentricity of the Earth's orbit, is one of the simplest. Another, having more important consequences, results from the changing direction of the axes of rotation in planets whose orbits are decidedly excentric. Every planet, during a certain long period, presents more of its northern than of its southern hemisphere to the sun at the time of its nearest approach to him; and then again, during a like period, presents more of its southern hemisphere than of its northern—a recurring coincidence which, though causing in some planets no sensible alterations of climate, involves in the case of the Earth an epoch of 21,000 years, during which each hemisphere goes through a cycle of temperate seasons, and seasons that are extreme in their heat and cold. Nor is this all. There is even a variation of this variation. For the summers and winters of the whole Earth become more or less strongly contrasted, as the excentricity of its orbit increases and decreases. Hence during increase of the excentricity, the epochs of moderately contrasted seasons and epochs of strongly contrasted seasons, through which alternately each hemisphere passes, must grow more and more different in the degrees of their contrasts; and contrariwise during decrease of the excentricity. So that in the quantity of light and heat which any portion of the Earth receives from the sun, there goes on a quadruple rhythm: that of day and night; that of summer and winter; that due to the changing position of the axis at perihelion and aphelion, taking 21,000 years to complete; and that involved by the variation of the orbit's excentricity, gone through in millions of years.

§ 95. Those terrestrial processes whose dependence on the solar heat is direct, of course exhibit a rhythm that corresponds to the periodically changing amount of heat which each part of the Earth receives. The simplest, though the least obtrusive, instance is supplied by the magnetic variations. In these there is a diurnal increase and decrease, an annual increase and decrease, and a decennial increase and decrease; the latter answering to a period during which the solar spots become alternately abundant and scarce: besides which known variations there are probably others corresponding with the astronomical cycles just described. More obvious examples are furnished by the movements of the ocean and the atmosphere. Marine currents from the equator to the poles above, and from the poles to the equator beneath, show us an unceasing backward and forward motion throughout this vast mass of water—a motion varying in amount according to the seasons, and compounded with smaller like motions of local origin. The similarly-caused general currents in the air, have similar annual variations similarly modified. Irregular as they are in detail, we still see in the monsoons and other tropical atmospheric disturbances, or even in our own equinoctial gales and spring east winds, a periodicity sufficiently decided. Again, we have an alternation of times during which evaporation predominates with times during which condensation predominates: shown in the tropics by strongly marked rainy seasons and seasons of drought, and in the temperate zones by corresponding changes of which the periodicity, though less definite, is still traceable. The diffusion and precipitation of water, besides the slow alternations answering to different parts of the year, furnish us with examples of rhythm of a more rapid kind. During wet weather, lasting, let us say, over some weeks, the tendency to condense, though greater than the tendency to evaporate, does not show itself in continuous rain; but the period is made up of rainy days and days that are wholly or partially fair. Nor is it in this rude alternation only that the law is manifested. During any day throughout this wet weather a minor rhythm is traceable; and especially so when the tendencies to evaporate and to condense are nearly balanced. Among mountains this minor rhythm and its causes may be studied to great advantage. Moist winds, which do not precipitate their contained water in passing over the comparatively warm lowlands, lose so much heat when they reach the cold mountain peaks, that condensation rapidly takes place. Water, however, in passing from the gaseous to the fluid state, gives out a considerable amount of heat; and hence the resulting clouds are warmer than the air that precipitates them, and much warmer than the high rocky surfaces round which they fold themselves. Hence in the course of the storm, these high rocky surfaces are raised in temperature, partly by radiation from the enwrapping cloud, partly by contact of the falling rain-drops. Giving off more heat than before, they no longer lower so greatly the temperature of the air passing over them; and so cease to precipitate its contained water. The clouds break; the sky begins to clear; and a gleam of sunshine promises that the day is going to be fine. But the small supply of heat which the cold mountain's sides have received, is soon lost: especially when the dispersion of the clouds permits free radiation into space. Very soon, therefore, these elevated surfaces, becoming as cold as at first, (or perhaps even colder in virtue of the evaporation set up,) begin again to condense the vapour in the air above; and there comes another storm, followed by the same effects as before. In lowland regions this action and reaction is usually less conspicuous, because the contrast of temperatures is less marked. Even here, however, it may be traced; and that not only on showery days, but on days of continuous rain; for in these we do not see uniformity: always there are fits of harder and gentler rain that are probably caused as above explained.

Of course these meteorologic rhythms involve something corresponding to them in the changes wrought by wind and water on the Earth's surface. Variations in the quantities of sediment brought down by rivers that rise and fall with the seasons, must cause variations in the resulting strata— alternations of colour or quality in the successive laminæ. Beds formed from the detritus of shores worn down and carried away by the waves, must similarly show periodic differences answering to the periodic winds of the locality. In so far as frost influences the rate of denudation, its recurrence is a factor in the rhythm of sedimentary deposits. And the geological changes produced by glaciers and icebergs must similarly have their alternating periods of greater and less intensity.

There is evidence also that modifications in the Earth's crust due to igneous action have a certain periodicity. Volcanic eruptions are not continuous but intermittent, and as far as the data enable us to judge, have a certain average rate of recurrence; which rate of recurrence is complicated by rising into epochs of greater activity and falling into epochs of comparative quiescence. So too is it with earthquakes and the elevations or depressions caused by them. At the mouth of the Mississippi, the alternation of strata gives decisive proof of successive sinkings of the surface, that have taken place at tolerably equal intervals. Everywhere, in the extensive groups of conformable strata that imply small subsidences recurring with a certain average frequency, we see a rhythm in the action and reaction between the Earth's crust and its molten contents—a rhythm compounded with those slower ones shown in the termination of groups of strata, and the commencement of other groups not conformable to them. There is even reason for suspecting a geological periodicity that is immensely slower and far wider in its effects; namely, an alternation of those vast upheavals and submergencies by which continents are produced where there were oceans, and oceans where there were continents. For supposing, as we may fairly do, that the Earth's crust is throughout of tolerably equal thickness, it is manifest that such portions of it as become most depressed below the average level, must have their inner surfaces most exposed to the currents of molten matter circulating within, and will therefore undergo a larger amount of what may be called igneous denudation; while, conversely, the withdrawal of the inner surfaces from these currents where the Earth's crust is most elevated, will cause a thickening more or less compensating the aqueous denudation going on externally. Hence those depressed areas over which the deepest oceans lie, being gradually thinned beneath and not covered by much sedimentary deposit above, will become areas of least resistance, and will then begin to yield to the upward pressure of the Earth's contents; whence will result, throughout such areas, long-continued elevations, ceasing only when the reverse state of things has been brought about. Whether this speculation be well or ill founded, does not however affect the general conclusion. Apart from it we have sufficient evidence that geologic processes are rhythmical.

§ 96. Perhaps nowhere are the illustrations of rhythm so numerous and so manifest as among the phenomena of life. Plants do not, indeed, usually show us any decided periodicities, save those determined by day and night and by the seasons. But in animals we have a great variety of movements in which the alternation of opposite extremes goes on with all degrees of rapidity. The swallowing of food is effected by a wave of constriction passing along the œsophagus; its digestion is accompanied by a muscular action of the stomach that is also undulatory; and the peristaltic motion of the intestines is of like nature. The blood obtained from this food is propelled not in a uniform current but in pulses; and it is aerated by lungs that alternately contract and expand. All locomotion results from oscillating movements: even where it is apparently continuous, as in many minute forms, the microscope proves the vibration of cilia to be the agency by which the creature is moved smoothly forwards.

Primary rhythms of the organic actions are compounded with secondary ones of longer duration. These various modes of activity have their recurring periods of increase and decrease. We see this in the periodic need for food, and in the periodic need for repose. Each meal induces a more rapid rhythmic action of the digestive organs; the pulsation of the heart is accelerated; and the inspirations become more frequent. During sleep, on the contrary, these several movements slacken. So that in the course of the twenty-four hours, those small undulations of which the different kinds of organic action are constituted, undergo one long wave of increase and decrease, complicated with several minor waves. Experiments have shown that there are still slower rises and falls of functional activity. Waste and assimilation are not balanced by every meal, but one or other maintains for some time a slight excess; so that a person in ordinary health is found to undergo an increase and decrease of weight during recurring intervals of tolerable equality. Besides these regular periods there are still longer and comparatively irregular ones; namely, those alternations of greater and less vigour, which even healthy people experience. So inevitable are these oscillations that even men in training cannot be kept stationary at their highest power, but when they have reached it begin to retrograde. Further evidence of rhythm in the vital movements is furnished by invalids. Sundry disorders are named from the intermittent character of their symptoms. Even where the periodicity is not very marked, it is mostly traceable. Patients rarely if ever get uniformly worse; and convalescents have usually their days of partial relapse or of less decided advance.

Aggregates of living creatures illustrate the general truth in other ways. If each species of organism be regarded as a whole, it displays two kinds of rhythm. Life as it exists in all the members of such species, is an extremely complex kind of movement, more or less distinct from the kinds of movement which constitutes life in other species. In each individual of the species, this extremely complex kind of movement begins, rises to its climax, declines, and ceases in death. And every successive generation thus exhibits a wave of that peculiar activity characterizing the species as a whole. The other form of rhythm is to be traced in that variation of number which each tribe of animals and plants is ever undergoing. Throughout the unceasing conflict between the tendency of a species to increase and the antagonistic tendencies, there is never an equilibrium: one always predominates. In the case even of a cultivated plant or domesticated animal, where artificial means are used to maintain the supply at a uniform level, we still see that oscillations of abundance and scarcity cannot be avoided. And among the creatures uncared for by man, such oscillations are usually more marked. After a race of organisms has been greatly thinned by enemies or lack of food, its surviving members become more favourably circumstanced than usual. During the decline in their numbers their food has grown relatively more abundant; while their enemies have diminished from want of prey. The conditions thus remain for some time favourable to their increase; and they multiply rapidly. By and by their food is rendered relatively scarce, at the same time that their enemies have become more numerous; and the destroying influences being thus in excess, their number begins to diminish again. Yet one more rhythm, extremely slow in its action, may be traced in the phenomena of Life, contemplated under their most general aspect. The researches of palæontologists show, that there have been going on, during the vast period of which our sedimentary rocks bear record, successive changes of organic forms. Species have appeared, become abundant, and then disappeared. Genera, at first constituted of but few species, have for a time gone on growing more multiform; and then have begun to decline in the number of their subdivisions; leaving at last but one or two representatives, or none at all. During longer epochs whole orders have thus arisen, culminated, and dwindled away. And even those wider divisions containing many orders have similarly undergone a gradual rise, a high tide, and a long-continued ebb. The stalked *Crinoidea*, for example, which, during the carboniferous epoch, became abundant, have almost disappeared: only a single species being extant. Once a large family of molluscs, the *Brachiopoda* have now become rare. The shelled Cephalopods, at one time dominant among the inhabitants of the ocean, both in number of forms and of individuals, are in our day nearly extinct. And after an "age of reptiles," there has come an age in which reptiles have been in great measure supplanted by mammals. Whether these vast rises and falls of different kinds of life ever undergo anything approaching to repetitions, (which they may possibly do in correspondence with those vast cycles of elevation and subsidence that produce continents and oceans,) it is sufficiently clear that Life on the Earth has not progressed uniformly, but in immense undulations.

§ 97. It is not manifest that the changes of consciousness are in any sense rhythmical. Yet here, too, analysis proves both that the mental state existing at any moment is not uniform, but is decomposable into rapid oscillations; and also that mental states pass through longer intervals of increasing and decreasing intensity.

Though while attending to any single sensation, or any group of related sensations constituting the consciousness of an object, we seem to remain for the time in a persistent and homogeneous condition of mind, a careful self-examination shows that this apparently unbroken mental state is in truth traversed by a number of minor states, in which various other sensations and perceptions are rapidly presented and disappear. From the admitted fact that thinking consists in the establishment of relations, it is a necessary corollary that the maintenance of consciousness in any one state to the entire exclusion of other states, would be a cessation of thought, that is, of consciousness. So that any seemingly continuous feeling, say of pressure, really consists of portions of that feeling perpetually recurring after the momentary intrusion of other feelings and ideas—quick thoughts concerning the place where it is felt, the external object producing it, its consequences, and other things suggested by association. Thus there is going on an extremely rapid departure from, and return to, that particular mental state which we regard as persistent. Besides the evidence of rhythm in consciousness which direct analysis thus affords, we may gather further evidence from the correlation between feeling and movement. Sensations and emotions expend themselves in producing muscular contractions. If a sensation or emotion were strictly continuous, there would be a continuous discharge along those motor nerves acted upon. But so far as experiments with artificial stimuli enable us to judge, a continuous discharge along the nerve leading to a muscle, does not contract it: a broken discharge is required—a rapid succession of shocks. Hence muscular contraction pre-supposes that rhythmic state of consciousness which direct observation discloses. A much more conspicuous rhythm, having longer waves, is seen during the outflow of emotion into dancing, poetry, and music. The current of mental energy that shows itself in these modes of bodily action, is not continuous, but falls into a succession of pulses. The measure of a dance is produced by the alternation of strong muscular contractions with weaker ones; and, save in measures of the simplest order such as are found among barbarians and children, this alternation is compounded with longer rises and falls in the degree of muscular excitement. Poetry is a form of speech which results when the emphasis is regularly recurrent; that is, when the muscular effort of pronunciation has definite periods of greater and less intensity—periods that are complicated with others of like nature answering to the successive verses. Music, in still more various ways, exemplifies the law. There are the recurring bars, in each of which there is a primary and a secondary beat. There is the alternate increase and decrease of muscular strain, implied by the ascents and descents to the higher and lower notes—ascents and descents composed of smaller waves, breaking the rises and falls of the larger ones, in a mode peculiar to each melody. And then we have, further, the alternation of *piano* and *forte* passages. That these several kinds of rhythm, characterizing æsthetic expression, are not, in the common sense of the word, artificial, but are intenser forms of an undulatory movement habitually generated by feeling in its bodily discharge, is shown by the fact that they are all traceable in ordinary speech; which in every sentence has its primary and secondary emphases, and its cadence containing a chief rise and fall complicated with subordinate rises and falls; and which is accompanied by a more or less oscillatory action of the limbs when the emotion is great. Still longer undulations may be observed by every one, in himself and in others, on occasions of extreme pleasure or extreme pain. Note, in the first place, that pain having its origin in bodily disorder, is nearly always perceptibly rhythmical. During hours in which it never actually ceases, it has its variations of intensity—fits or paroxysms; and then after these hours of suffering there usually come hours of comparative ease. Moral pain has the like smaller and larger waves. One possessed by intense grief does not utter continuous moans, or shed tears with an equable rapidity; but these signs of passion come in recurring bursts. Then after a time during which such stronger and weaker waves of emotion alternate, there comes a calm—a time of comparative deadness; to which again succeeds another interval, when dull sorrow rises afresh into acute anguish, with its series of paroxysms. Similarly in great delight, especially as manifested by children who have its display less under control, there are visible variations in the intensity of feeling shown—fits of laughter and dancing about, separated by pauses in which smiles, and other slight manifestations of pleasure, suffice to discharge the lessened excitement. Nor are there wanting evidences of mental undulations greater in length than any of these—undulations which take weeks, or months, or years, to complete themselves. We continually hear of moods which recur at intervals. Very many persons have their epochs of vivacity and depression. There are periods of industry following periods of idleness; and times at which particular subjects or tastes are cultivated with zeal, alternating with times at which they are neglected. Respecting which slow oscillations, the only qualification to be made is, that being affected by numerous influences, they are comparatively irregular.

§ 98. In nomadic societies the changes of place, determined as they usually are by exhaustion or failure of the supply of food, are periodic; and in many cases show a recurrence answering to the seasons. Each tribe that has become in some degree fixed in its locality, goes on increasing, till under the pressures of unsatisfied desires, there results migration of some part of it to a new region—a process repeated at intervals. From such excesses of population, and such successive waves of migration, come conflicts with other tribes; which are also increasing and tending to diffuse themselves. This antagonism, like all others, results not in an uniform motion, but in an intermittent one. War, exhaustion, recoil—peace, prosperity, and renewed aggression:—see here the alternation more or less discernible in the military activities of both savage and civilized nations. And irregular as is this rhythm, it is not more so than the different sizes of the societies, and the extremely involved causes of variation in their strengths, would lead us to anticipate.

Passing from external to internal changes, we meet with this backward and forward movement under many forms. In the currents of commerce it is especially conspicuous. Exchange during early times is almost wholly carried on at fairs, held at long intervals in the chief centres of population. The flux and reflux of people and commodities which each of these exhibits, becomes more frequent as national development leads to greater social activity. The more rapid rhythm of weekly markets begins to supersede the slow rhythm of fairs. And eventually the process of exchange becomes at certain places so active, as to bring about daily meetings of buyers and sellers—a daily wave of accumulation and distribution of cotton, or corn, or capital. If from exchange we turn to production and consumption, we see undulations, much longer indeed in their periods, but almost equally obvious. Supply and demand are never completely adapted to each other; but each of them from time to time in excess, leads presently to an excess of the other. Farmers who have one season produced wheat very abundantly, are disgusted with the consequent low price; and next season, sowing a much smaller quantity, bring to market a deficient crop; whence follows a converse effect. Consumption undergoes parallel undulations that need not be specified. The balancing of supplies between different districts, too, entails analogous oscillations. A place at which some necessary of life is scarce, becomes a place to which currents of it are set up from other places where it is relatively abundant; and these currents from all sides lead to a wave of accumulation where they meet—a glut: whence follows a recoil—a partial return of the currents. But the undulatory character of these actions is perhaps best seen in the rises and falls of prices. These, given in numerical measures which may be tabulated and reduced to diagrams, show us in the clearest manner how commercial movements are compounded of oscillations of various magnitudes. The price of consols or the price of wheat, as thus represented, is seen to undergo vast ascents and descents whose highest and lowest points are reached only in the course of years. These largest waves of variation are broken by others extending over periods of perhaps many months. On these again come others having a week or two's duration. And were the changes marked in greater detail, we should have the smaller undulations that take place each day, and the still smaller ones which brokers telegraph from hour to hour. The whole outline would show a complication like that of a vast ocean-swell, on whose surface there rise large billows, which themselves bear waves of moderate size, covered by wavelets, that are roughened by a minute ripple. Similar diagrammatic representations of births, marriages, and deaths, of disease, of crime, of pauperism, exhibit involved conflicts of rhythmical motions throughout society under these several aspects.

There are like characteristics in social changes of a more complex kind. Both in England and among continental nations, the action and reaction of political progress have come to be generally recognized. Religion, besides its occasional revivals of smaller magnitude, has its long periods of exaltation and depression—generations of belief and self-mortification, following generations of indifference and laxity. There are poetical epochs, and epochs in which the sense of the beautiful seems almost dormant. Philosophy, after having been awhile predominant, lapses for a long season into neglect; and then again slowly revives. Each science has its eras of deductive reasoning, and its eras when attention is chiefly directed to collecting and colligating facts. And how in such minor but more obtrusive phenomena as those of fashion, there are ever going on oscillations from one extreme to the other, is a trite observation.

As may be foreseen, social rhythms well illustrate the irregularity that results from combination of many causes. Where the variations are those of one simple element in national life, as the supply of a particular commodity, we do indeed witness a return, after many involved movements, to a previous condition—the price may become what it was before: implying a like relative abundance. But where the action is one into which many factors enter, there is never a recurrence of exactly the same state. A political reaction never brings round just the old form of things. The rationalism of the present day differs widely from the rationalism of the last century. And though fashion from time to time revives extinct types of dress, these always re-appear with decided modifications.

§ 99. The universality of this principle suggests a question like that raised in foregoing cases. Rhythm being manifested in all forms of movement, we have reason to suspect that it is determined by some primordial condition to action in general. The tacit implication is that it is deducible from the persistence of force. This we shall find to be the fact.

When the prong of a tuning-fork is pulled on one side by the finger, a certain extra tension is produced among its cohering particles; which resist any force that draws them out of their state of equilibrium. As much force as the finger exerts in pulling the prong aside, so much opposing force is brought into play among the cohering particles. Hence, when the prong is liberated, it is urged back by a force equal to that used in deflecting it. When, therefore, the prong reaches its original position, the force impressed on it during its recoil, has generated in it a corresponding amount of momentum—an amount of momentum nearly equivalent, that is, to the force originally impressed (nearly, we must say, because a certain portion has gone in communicating motion to the air, and a certain other portion has been transformed into heat). This momentum carries the prong beyond the position of rest, nearly as far as it was originally drawn in the reverse direction; until at length, being gradually used up in producing an opposing tension among the particles, it is all lost. The opposing tension into which the expended momentum has been transformed, then generates a second recoil; and so on continually—the vibration eventually ceasing only because at each movement a certain amount of force goes in creating atmospheric and etherial undulations. Now it needs but to contemplate this repeated action and reaction, to see that it is, like every action and reaction, a consequence of the persistence of force. The force exerted by the finger in bending the prong cannot disappear. Under what form then does it exist? It exists under the form of that cohesive tension which it has generated among the particles. This cohesive tension cannot cease without an equivalent result. What is its equivalent result? The momentum generated in the prong while being carried back to its position of rest. This momentum too—what becomes of it? It must either continue as momentum, or produce some correlative force of equal amount. It cannot continue as momentum, since change of place is resisted by the cohesion of the parts; and thus it gradually disappears by being transformed into tension among these parts. This is re-transformed into the equivalent momentum; and so on continuously. If instead of motion that is directly antagonized by the cohesion of matter, we consider motion through space, the same truth presents itself under another form. Though here no opposing force seems at work, and therefore no cause of rhythm is apparent, yet its own accumulated momentum must eventually carry the moving body beyond the body attracting it; and so must become a force at variance with that which generated it. From this conflict, rhythm necessarily results as in the foregoing case. The force embodied as momentum in a given direction, cannot be destroyed; and if it eventually disappears, it re-appears in the reaction on the retarding body; which begins afresh to draw the now arrested mass back from its aphelion. The only conditions under which there could be absence of rhythm—the only conditions, that is, under which there could be a continuous motion through space in the same straight line for ever, would be the existence of an infinity void of everything but the moving body. And neither of these conditions can be represented in thought. Infinity is inconceivable; and so also is a motion which never had a commencement in some pre-existing source of power. Thus, then, rhythm is a necessary characteristic of all motion. Given the coexistence everywhere of antagonist forces—a postulate which, as we have seen, is necessitated by the form of our experience—and rhythm is an inevitable corollary from the persistence of force.

Hence, throughout that re-arrangement of parts which constitutes Evolution, we must nowhere expect to see the change from one position of things to another, effected by continuous movement in the same direction. Be it in that kind of Evolution which the inorganic creation presents, or in that presented by the organic creation, we shall everywhere find a periodicity of action and reaction—a backward and forward motion, of which progress is a differential result.

15. After having for some years supposed myself alone in the belief that all motion is rhythmical, I discovered that my friend Professor Tyndall also held this doctrine.

CHAPTER XII
THE CONDITIONS ESSENTIAL TO EVOLUTION

§ 100. One more preliminary is needful before proceeding. We have still to study the conditions under which alone, Evolution can take place.

The process to be interpreted is, as already said, a certain change in the arrangement of parts. That increase of heterogeneity commonly displayed throughout Evolution, is not an increase in the number of kinds of ultimate or undecomposable units which an aggregate contains; but it is a change in the distribution of such units. If it be assumed that what we call chemical elements, are absolutely simple (which is, however, an hypothesis having no better warrant than the opposite one); then it must be admitted that in respect to the number of kinds of matter contained in it, the Earth is not more heterogeneous at present than it was at first—that in this respect, it would be as heterogeneous were all its undecomposable parts uniformly mixed, as it is now, when they are arranged and combined in countless different ways. But the increase of heterogeneity with which we have to deal, and of which alone our senses can take cognizance, is that produced by the passage from unity of distribution to variety of distribution. Given an aggregate consisting of several orders of primitive units that are unchangeable; then, these units may be so uniformly dispersed among each other, that any portion of the mass shall be like any other portion in its sensible properties; or they may be so segregated, simply and in endless combinations, that the various portions of the mass shall not be like each other in their sensible properties. A transformation of one of these arrangements into the other, is that which constitutes Evolution. We have to analyze the process through which structural uniformity becomes structural multiformity—to ascertain how the originally equal relations of position among the mixed units, pass into relations of position that are more and more unequal, and more and more numerous in their kinds of inequality; and how this takes place throughout all the ascending grades of compound units, until we come even to those of which societies are made up.

Change in the relations of position among the component units, simple or complex, being the phenomenon we have to interpret; we must first inquire what are the circumstances which prevent its occurrence, and what are the circumstances which facilitate it.

§ 101. The constituents of an aggregate cannot be re-arranged, unless they are moveable: manifestly, they must not be so firmly bound together that the incident force fails to alter their positions. No bodies are, indeed, possessed of this absolute rigidity; since an incident force in being propagated through a body, always produces temporary alterations in the relative positions of its units, if not permanent alterations. It is true also, that even permanent re-arrangements of the units may be thus wrought throughout the interiors of comparatively dense masses, without any outward sign: as happens with certain crystals, which, on exposure to sunlight, undergo molecular changes so great as to alter their planes of cleavage. Nevertheless, since total immobility of the parts must totally negative their re-arrangement; and since that comparative immobility which we see in very coherent matter, is a great obstacle to re-arrangement; it is self-evident that Evolution can be exhibited in any considerable degree, only where there is comparative mobility of parts. On the other hand, those definite distributive changes which constitute Evolution, cannot be extensively or variously displayed, where the mobility of the parts is extreme. In liquids, the cohesion of the units is so slight that there is no permanency in their relations of position to each other. Such re-arrangement as any incident force generates, is immediately destroyed again by the momentum of the constituents moved; and so, nothing but that temporary heterogeneity seen in circulating currents, can be produced. The like still more obviously holds of gases. Thus, while the theoretical limits between which Evolution is possible, are absolute immobility of parts and absolute mobility of parts; we may say that practically, Evolution cannot go on to any considerable extent where the mobility is very great or very little. A few examples will facilitate the realization of this truth.

The highest degrees of Evolution are found in semi-solid bodies, or bodies that come midway between the two extremes specified. Even semi-solid bodies of the inorganic class, exhibit the segregation of mixed units with comparative readiness: witness the fact to which attention was first drawn by Mr. Babbage, that when the pasty mixture of ground flints and kaolin, prepared for the manufacture of porcelain, is kept some time, it becomes gritty and unfit for use, in consequence of the particles of silica separating themselves from the rest, and uniting together in grains; or witness the fact known to every housewife, that in long-kept currant-jelly the sugar takes the shape of imbedded crystals. While throughout the immense majority of the semi-solid bodies, namely, the organic bodies, the proclivity to a re-arrangement of parts is so comparatively great, as to be usually taken for a distinctive characteristic of them. Among organic bodies themselves, we may trace contrasts having a like significance. It is an accepted generalization that, other things equal, the rate of Evolution is greatest where the plasticity is most marked. In that portion of an egg which displays the formative processes during the early stages of incubation, the changes of arrangement are more rapid than those which an equal portion of the body of a hatched chick undergoes. As may be inferred from their respective powers to acquire habits and aptitudes, the structural modifiability of a child is greater than that of an adult man; and the structural modifiability of an adult man is greater than that of an old man: contrasts which are accompanied by corresponding contrasts in the densities of the tissues; since the ratio of water to solid matter diminishes with advancing age. The most decisive proof, however, is furnished by those marked retardations or arrests of organic change, that take place when the tissues suffer a great loss of water. Certain of the lower animals, as the *Rotifera*, may be rendered apparently lifeless by desiccation, and will yet revive when wetted: as their substance passes from the fluid-solid to the solid state, it ceases to be the seat of those changes which constitute functional activity and cause structural advance; and such changes recommence as their substance passes from the solid to the fluid-solid state. Analogous instances occur among much higher animals. When the African rivers which it inhabits are dried up, the *Lepidosiren* remains torpid in the hardened mud, until the return of the rainy season brings water. Humboldt states that during the summer drought, the alligators of the Pampas lie buried in a state of suspended animation beneath the parched surface, and struggle up out of the earth as soon as it becomes humid. Now though we have no proof that these partial arrests of vital activity, are consequent on the reduction of the fluid-solid tissues to a more solid form; yet their occurrence along with a cessation in the supply of water, is reason for suspecting that this is the case. And similarly, though in the more numerous instances where loss of water leads to complete arrest of vital activity, we are unable to say that the immediate cause is a stoppage of molecular changes that results from a diminution of molecular mobility; yet it seems not improbable that this is the rationale of death by thirst.

Probably few will expect to find this same condition to Evolution, illustrated in aggregates so widely different in kind as societies. Yet even here it may be shown that no considerable degree of Evolution is exhibited, where there is either great mobility of the parts, or great immobility of them. In such tribes as those inhabiting Australia, we see extremely little cohesion among the units: there is neither that partial fixity of relative positions which results from the commencement of agriculture, nor that partial fixity of relative positions implied by the establishment of social grades. And along with this want of cohesion, we find an absence of permanent differentiations. Conversely, in societies of the oriental type, where accumulated traditions, laws, and usages, and long-fixed class-arrangements, exercise great restraining power over individual actions, we find Evolution almost stopped. Through the medium of institutions and opinions, the forces brought to bear on each unit by the rest, are so great as to prevent the units from sensibly yielding to forces tending to re-arrange them. The condition most favourable to increase of social heterogeneity, is a medium coherence among the parts—a moderate facility of change in the relations of citizens, joined with a moderate resistance to such change—a considerable freedom of individual actions, qualified by a considerable restraint over individual actions—a certain attachment to pre-established arrangements, and a certain readiness to be impelled by new influences into new arrangements—a compromise between fixity and unfixity such as that which we, perhaps as much as any nation, exhibit.

§ 102. Another condition to Evolution, of the same order as the last though of a different genus, must be noted. We have found that permanent re-arrangement among the units of an aggregate, can take place only when they have neither extreme immobility nor extreme mobility. The mobility and immobility thus far considered (at least in all aggregates except social ones) are those due to mechanical cohesion. There is, however, what we must call chemical cohesion, which also influences the mobility of the units, and consequently the re-arrangement of them. Manifestly, if two or more kinds of units contained in any aggregate, are united by powerful affinities, an incident force, failing to destroy their cohesions, will not cause such various re-arrangements as it would, could it produce new chemical combinations as well as new mechanical adjustments. On the other hand, chemical affinities that are easily overcome, must be favourable to multiplied re-arrangements of the units.

This condition, as well as the preceding one, is fulfilled in the highest degree, by those aggregates which most variously display the transformation of the uniform into the multiform. Organic bodies are on the average distinguished from inorganic bodies, by the readiness with which the compounds they consist of undergo decomposition, and recomposition: the chemical cohesions of their components are so comparatively small, that small incident forces suffice to overcome them and cause transpositions of the components. Further, between the two great divisions of organisms, we find a contrast in the degree of Evolution co-existing with a contrast in the degree of chemical modifiability. As a class, the nitrogenous compounds are peculiarly unstable; and, speaking generally, these are present in much larger quantities in animal tissues than they are in vegetal tissues; while, speaking generally, animals are much more heterogeneous than plants.

Under this head it may be well also to point out that, other things equal, the structural variety which is possible in any aggregate, must bear a relation to the number of kinds of units contained in the aggregate. A body made up of units of one order, cannot admit of so many different re-arrangements, as one made up of units of two orders. And each additional order of units must increase, in a geometrical proportion, the number of re-arrangements that may be made.

§ 103. Yet one more condition to be specified, is the state of agitation in which the constituents of an aggregate are kept. A familiar expedience will introduce us to this condition. When a vessel has been filled to the brim with loose fragments, shaking the vessel causes them to settle down into less space, so that more may be put in. And when among these fragments, there are some of much greater specific gravity than the rest, these will, in the course of a prolonged shaking, find their way to the bottom. What now is the meaning of these two results, when expressed in general terms? We have a group of units acted on by an incident force—the attraction of the Earth. So long as these units are not agitated, this incident force produces no changes in their relative positions; agitate them, and immediately their loose arrangement passes into a more compact arrangement. Again, so long as they are not agitated, the incident force cannot separate the heavier units from the lighter; agitate them, and immediately the heavier units begin to segregate. By these illustrations, a rude idea will be conveyed of the effect which vibration has in facilitating those re-arrangements which constitute Evolution. What here happens with visible units subject to visible oscillations, happens also with invisible units subject to invisible oscillations.

One or two cases in which these oscillations are of mechanical origin, may first be noted. When a bar of steel is suspended in the magnetic meridian, and repeatedly so struck as to send vibrations through it, it becomes magnetized: the magnetic force of the Earth, which does not permanently affect it while undisturbed, alters its internal state when a mechanical agitation is propagated among its particles; and the alteration is believed by physicists, to be a molecular re-arrangement. It may be fairly objected that this re-arrangement is hypothetical; and did the fact stand alone, it would be of little worth. It gains significance, however, when joined with the fact that in the same substance, long-continued mechanical vibrations are followed by molecular re-arrangements that are abundantly visible. A piece of iron which, when it leaves the workshop, is fibrous in structure, will become crystalline if exposed to a perpetual jar. Though the polar forces mutually exercised by the atoms, fail to change their disorderly arrangement into an orderly arrangement while the atoms are relatively quiescent, these forces produce this change when the atoms are kept in a state of intestine disturbance.

But the effects which visible oscillations and oscillations sensible to touch, have in facilitating the re-arrangement of parts by an incident force, are insignificant compared with the effects which insensible oscillations have in aiding such change of structure. It is a doctrine now generally accepted among men of science, that the particles of tangible matter, as well as the particles of ether, undulate. As interpreted in conformity with this doctrine, the heat of a body is simply its state of molecular motion. A mass which feels cold, is one having but slight molecular motion, and conveying but slight molecular motion to the surrounding medium or to the hand touching it. A mass hot enough to radiate a sensible warmth, is one of which the more violently agitated molecules, communicate increased undulations to the surrounding ethereal medium; while the burn inflicted by it on the skin, is the expression of increased undulations of the organic molecules. Such further heat as produces softening and a consequent distortion of the mass, is an agitation so much augmented that the units can no longer completely maintain their relative positions. Fusion is an agitation so extreme, that the relative positions of the units are changeable with ease. When, finally, at a still higher temperature, the liquid is transformed into a gas, the explanation is, that the oscillations are so violent as to overbalance that force which held the units in close contiguity—so violent as to keep the units at those relatively great distances apart to which they are now thrown. Since the establishment of the correlation between heat and motion first gave probability to this hypothesis, it has been receiving various confirmations— especially by recent remarkable discoveries respecting the absorption of heat by gases. Prof. Tyndall has proved that the quantity of heat which any gas takes up from rays of heat passing through it, has a distinct relation to the complexity of the atoms composing the gas. The simple gases abstract but little; the gases composed of binary atoms abstract, say in round numbers, a hundred times as much; while the gases composed of atoms severally containing three, four, or more simple ones, abstract something like a thousand times as much. These differences Prof. Tyndall regards as due to the different abilities of the different atoms to take up, in the increase of their own undulations, those undulations of the ethereal medium which constitute heat—an interpretation in perfect accordance with the late results of spectrum-analysis; which go to show that the various elementary atoms, when in an aeriform state, intercept those luminiferous vibrations of the ether which are in unison or harmony with their own. And since it holds of solid as of gaseous matters, that those consisting of simple units transmit heat far more readily than those consisting of complex units; we get confirmation of the inference otherwise reached, that the units of matter in whatever state of aggregation they exist, oscillate, and that variations of temperature are variations in the amounts of their oscillations.

Proceeding on this hypothesis, which it would be out of place here to defend at greater length, we have now to note how the re-arrangement of parts is facilitated by these insensible vibrations, as we have seen it to be by sensible vibrations. One or two cases of physical re-arrangement may first be noted. When some molten glass is dropped into water, and when its outside is thus, by sudden solidification, prevented from partaking in that contraction which the subsequent cooling of the inside tends to produce; the units are left in such a state of tension, that the mass flies into fragments if a small portion of it be broken off. But now, if this mass be kept for a day or two at a considerable heat, though a heat not sufficient to alter its form or produce any sensible diminution of hardness, this extreme brittleness disappears: the component particles being thrown into greater agitation, the tensile forces are enabled to re-arrange them into a state of equilibrium. An illustration of another order is furnished by the subsidence of fine precipitates. These sink down very slowly from solutions that are cold; while warm solutions deposit them with comparative rapidity. That is to say, an increase of molecular vibration throughout the mass, allows the suspended particles to separate more readily from the particles of fluid. The effect of heat on chemical re-arrangement is so familiar, that examples are scarcely needed. Be the substances concerned gaseous, liquid, or solid, it equally holds that their chemical unions and disunions are aided by a rise of temperature. Affinities which do not suffice to effect the re-arrangement of mixed units that are in a state of feeble agitation, suffice to effect it when the agitation is raised to a certain point. And so long as this molecular motion is not great enough to prevent those chemical cohesions which the affinities tend to produce, increase of it gives increased facility of chemical re-arrangement.

This condition, in common with the preceding ones, is fulfilled most completely in those aggregates which exhibit the phenomena of Evolution in the highest degree; namely, the organic aggregates. And throughout the various orders and states of these, we find minor contrasts showing the relation between amount of molecular vibration and activity of the metamorphic changes. Such contrasts may be arranged in the several following groups. Speaking generally, the phenomena of Evolution are manifested in a much lower degree throughout the vegetal kingdom than throughout the animal kingdom; and speaking generally, the heat of plants is less than that of animals. Among plants themselves, the organic changes vary in rate as the temperature varies. Though light is the agent which effects those molecular changes causing vegetal growth, yet we see that in the absence of heat, such changes are not effected: in winter there is enough light, but the heat being insufficient, plant-life is suspended. That this is the sole cause of the suspension, is proved by the fact that at the same season, plants contained in hot-houses, where they receive even a smaller amount of light, go on producing leaves and flowers. A comparison of the several divisions of the animal kingdom with each other, shows among them parallel relations. Regarded as a whole, vertebrate animals are higher in temperature than invertebrate ones; and they are as a whole higher in organic activity and development. Between subdivisions of the vertebrata themselves, like differences in the state of molecular vibration, accompany like differences in the degree of evolution. The least heterogeneous of the vertebrata are the fishes; and in most cases, the heat of fishes is nearly the same as that of the water in which they swim: only some of them being decidedly warmer. Though we habitually speak of reptiles as cold-blooded; and though they have not much more power than fishes of maintaining a temperature above that of their medium; yet since their medium (which is, in the majority of cases, the air of warm climates) is on the average warmer than the medium inhabited by fishes, the temperature of the class of reptiles is higher than that of the class of fishes; and we see in them a correspondingly higher complexity. The much more active molecular agitation in mammals and birds, is associated with a considerably greater multiformity of structure and a very much greater vivacity. And though birds, which are hotter blooded than mammals, do not show us a greater multiformity; yet, judging from their apparently greater locomotive powers, we may infer more rapid functional changes, which, equally with structural changes, imply molecular re-arrangement. The most instructive contrasts, however, are those presented by the same organic aggregates at different temperatures. Thus we see that ova undergoing development, must be kept more or less warm—that in the absence of a certain molecular vibration, the re-arrangement of parts does not go on. We see, again, that in hybernating animals, loss of heat carried to a particular point, results in extreme retardation of the organic changes. Yet further, we see that in animals which do not hybernate, as in man, prolonged exposure to extreme cold, produces an irresistible tendency to sleep (which implies a lowering of the functional activity); and then, if the abstraction of heat continues, this sleep ends in death, or arrest of functional activity. Lastly, we see that when the temperature is lowered till the contained water solidifies, there is a stoppage not only of those molecular re-arrangements which constitute life and development, but also of those molecular re-arrangements which constitute decomposition.

Evidently then, both sensible and insensible agitations among the components of an aggregate, facilitate any re-distributions to which there may be a tendency. When that rhythmic change in the relative positions of the units which constitutes vibration, is considerable, the relative positions of the units more readily undergo permanent changes through the action of incident forces.

§ 104. These special conditions to Evolution, are clearly but different forms of one general condition. The abstract proposition, that a permanent re-arrangement of units is possible only when they have neither absolute immobility nor absolute mobility with respect to each other, we saw to be practically equivalent to the proposition, that extreme cohesion and extreme want of cohesion among the units are unfavourable to Evolution. Be this cohesion or want of cohesion that which physically characterizes the matter as we ordinarily know it; be it that cohesion or want of cohesion distinguished as chemical; or be it that cohesion or want of cohesion consequent on the degree of molecular vibration; matters not, in so far as the general conclusion is concerned. Inductively as well as deductively, we find that the genesis of such permanent changes in the relative positions of parts, as can be effected without destroying the continuity of the aggregate, implies a medium stability in the relative positions of the parts: be this stability physical, chemical, or that which varies with the state of agitation. And as might be anticipated *à priori*, it is proved *à posteriori*, that this re-arrangement of parts goes on most actively in those aggregates whose units are moderately influenced by all these forces which affect their mobility.

Here also may properly be added the remark, that to effect these changes in the relative positions of parts, the incident forces must range within certain limits. It is wholly a question of the ratio between those agencies which hold the units in their positions, and those agencies which tend to change their positions. Having given intensities in the powers that oppose re-arrangement, there need proportionate intensities in the powers that work re-arrangement. As there must be neither too great nor too little cohesion; so there must be neither too little nor too great amounts of the influences antagonistic to cohesion. While a slight mechanical strain produces no lasting alterations in the relative positions of parts, an excessive mechanical strain causes disruption—causes so great an alteration in the relative positions of parts as to destroy their union in one aggregate. While a very feeble chemical affinity brought to bear on the associated units, fails to work any re-arrangement of them; a chemical affinity that is extremely intense, destroys their structural continuity, and reduces such complex re-arrangements as have been made, to comparatively simple ones. And while in the absence of adequate thermal undulations, the units have not freedom enough to obey the re-arranging influences impressed on them, the incidence of violent thermal undulations gives them such extreme freedom that they break their connexions, and the aggregate lapses into a liquid or gaseous form.

On the one hand, therefore, the statical forces which uphold the state of aggregation must not be so great as wholly to prevent those changes of relative position among the units which the dynamical forces tend to produce; and, on the other hand, the dynamical forces must not be so great as wholly to overcome the statical forces, and destroy the state of aggregation. The excess of the dynamical forces must be sufficient to produce Evolution, but not sufficient to produce Dissolution.

§ 105. And now we are naturally introduced to a consideration which, though it does not come quite within the limits of this chapter as expressed in its title, may yet be more conveniently dealt with here than elsewhere. Hitherto we have studied the metamorphosis of things, only as exhibited in the changed distribution of matter. It remains to look at it as exhibited in the changed distribution of motion. The definition of Evolution in its material aspect, has to be supplemented by a definition of Evolution in its dynamical aspect.

On inquiring the source of the sensible motions seen in every kind of Evolution, we find them all traceable to insensible motions; either of that tangible matter which we perceive as constituting the objects around us, or of that intangible matter which we infer as occupying space. A brief reconsideration of the facts will make this obvious. The formation of celestial bodies, supposing it caused by the union of dispersed units, must, from the beginning, have involved a diminished motion of these units with respect to each other; and such motion as each resulting body acquired, must previously have existed in the motions of its units. If concrete matter has arisen by the aggregation of diffused matter, then concrete motion has arisen by the aggregation of diffused motion. That which now exists as the movement of masses, implies the cessation of an equivalent molecular movement. Those transpositions of matter which constitute geological changes, are clearly referable to the same source. As before shown, the denudation of lands and deposit of new strata, are effected by water in the course of its descent from the clouds to the sea, or during the arrest of those undulations produced on it by winds; and, as before shown, the elevation of water to the height whence it fell, is due to solar heat, as is also the genesis of those aerial currents which drift it about when evaporated and agitate its surface when condensed. That is to say, the molecular motion of the etherial medium, is transformed into the motion of gases, thence into the motion of liquids, and thence into the motion of solids—stages in each of which, successively, a certain amount of molecular motion is lost and an equivalent motion of masses produced. If we seek the origin of vital movements, we soon reach a like conclusion. The actinic rays issuing from the Sun, enable the plant to reduce special elements existing in gaseous combination around it, to a solid form,—enable the plant, that is, to grow and carry on its functional changes. And since growth, equally with circulation of sap, is a mode of sensible motion, while those rays which have been expended in generating it consist of insensible motions, we have here, too, a transformation of the kind alleged. Animals, derived as their forces are, directly or indirectly, from plants, carry this transformation a step further. The automatic movements of the viscera, together with the voluntary movements of the limbs and body at large, arise at the expense of certain molecular movements throughout the nervous and muscular tissues; and these originally arose at the expense of certain other molecular movements propagated by the Sun to the Earth; so that both the structural and functional motions which organic Evolution displays, are motions of aggregates generated by the arrested motions of units. Even with the aggregates of these aggregates the same rule holds. For among associated men, the progress is ever towards a merging of individual actions in the actions of corporate bodies. An undeveloped society is composed of members between whom there is little concert: they fulfil their several wants without mutual aid; and only on occasions of aggression or defence, act together—occasions on which their combination, small as it is in extent, frequently fails because it is so imperfect. In the course of civilization, however, co-operation becomes step by step more decided. As tribes grow into nations, there result larger aggregates, each of which has a joint political life—a common policy and movement with respect to other aggregates. Legislative and administrative progress, involves an increase in the number of restraining agents brought into united and simultaneous action. In military organization, we see an advance from small undisciplined hordes of armed men, to vast bodies of regular troops, so drilled that the movements of the units are entirely subordinated to the movements of the masses. Nor does industrial development fail to show parallel changes. Beginning with independent workers, and passing step by step to the employment of several assistants by one master, there has ever been, and still is, a progress towards the co-operation of greater masses of labourers in the same establishment, and towards the union of capitalists into more numerous and larger companies: in both which kinds of combined action, equivalent amounts of individual action disappear. Under all its forms, then, Evolution, considered dynamically, is a decrease in the relative movements of parts, and an increase in the relative movements of wholes—using the words parts and wholes in their widest senses. From the infinitesimal motions of those infinitesimal units composing the etherial medium, to the larger though still insensible motions of the larger though still insensible units composing gaseous, fluid, and solid matter, and thence to the visible motions of visible aggregates, the advance is from molecular motion to the motion of masses.

But now what of the converse process? If the foregoing proposition is true, then a change from the motion of masses to molecular motion, is the opposite to Evolution—is Dissolution. Is this so? Of inorganic dissolution we have but little experience; or at least, our experience of it is on too small a scale to exhibit it as the antithesis of Evolution. We know, indeed, that when solids are dissolved in liquids, their dissolution implies increased movements of their units, at the expense of diminished movements among the units of their solvents; and we know that when a liquid evaporates, its dissipation or dissolution similarly implies greater relative movements of the units, and decrease of such combined movement as they before had. But since these small aggregates of inorganic matter, do not exhibit the phenomena of Evolution, save in the form of simple integration; so they do not exhibit the phenomena of dissolution, save in the form of simple disintegration. Of organic dissolution, however, our experience suffices to show that it is a decrease of combined motion, and an increase in the motion of uncombined parts. The gradual cessation of functions, vegetal or animal, is a cessation of the sensible movements of fluids and solids. In animals, the impulsions of the body from place to place, first cease; presently the limbs cannot be stirred; later still the respiratory actions stop; finally the heart becomes stationary, and, with it, the circulating fluids. That is, the transformation of molecular motion into the motion of masses, comes to an end. What next takes place? We cannot say that sensible movements are transformed into insensible movements; for sensible movements no longer exist. Nevertheless, the process of decay involves an increase of insensible movements; since this is far greater in the gases generated by decomposition, than it is in the fluid-solid matters generating them. Indeed, it might be contended that as, during Dissolution, there is a change from the vibration of large compound atoms to the vibration of small and comparatively simple ones, the process is strictly antithetical to that of Evolution. In conformity with the now current conception lately explained, each of the highly complex chemical units composing an organic body, possesses a rhythmic movement—a movement in which its many component units jointly partake. When decomposition breaks up these highly complex atoms, and their constituents assume a gaseous form, there is both an increase of molecular motion implied by the diffusion, and a further increase implied by the resolving of such motions as the aggregate atoms possessed, into motions of their constituent atoms. So that in organic dissolution we have, first, an end put to that transformation of the motion of units into the motion of aggregates, which constitutes Evolution, dynamically considered; and we have also, though in a subtler sense, a transformation of the motion of aggregates into the motion of units. The formula equally applies to the dissolution of a society. When social ties, be they governmental or industrial, are destroyed, the combined actions of citizens lapse into uncombined actions. Those general forces which restrained individual doings, having disappeared, the only remaining restraints are those separately exercised by individuals on each other. There are no longer any of the joint operations by which men satisfy their wants; and, in so far as they can, they satisfy their wants by separate operations. That is to say, the movement of parts replaces the movement of wholes.

Under its dynamical aspect then, Evolution, so far as we can trace it, is a change from molecular motion to the motion of masses; while Dissolution, so far as we can trace it, is a change from the motion of masses to molecular motion.

§ 106. To these abstract definitions may be added concrete ones. Besides an integration of motions corresponding to the integration of masses, Evolution involves an increase in the multiformity of the motions, corresponding to the increase in the multiformity of the masses. If, contemplating it as materially displayed, we find Evolution to consist in the change from an indefinite, homogeneous distribution of parts to a definite, heterogeneous distribution of parts; then, contemplating Evolution as dynamically displayed, it consists in a change from indefinite, homogeneous motions to definite, heterogeneous motions.

This change takes place under the form of an increased variety of rhythms. We have already seen that all motion is rhythmical, from the infinitesimal vibrations of infinitesimal molecules, up to those vast oscillations between perihelion and aphelion performed by vast celestial bodies. And as the contrast between these extreme cases suggests, a multiplication of rhythms must accompany a multiplication in the degrees and modes of aggregation, and in the relations of the aggregated masses to incident forces. The degree or mode of aggregation will not, indeed, affect the rate or extent of rhythm where the incident force increases as the aggregate increases, which is the case with gravitation: here the only cause of variation in rhythm, is difference of relation to the incident forces; as we see in a pendulum, which, though unaffected in its movements by a change in the weight of the bob, alters its rate of oscillation when taken to the equator. But in all cases where the incident forces do not vary as the masses, every new order of aggregation initiates a new order of rhythm: witness the conclusion drawn from the recent researches into radiant heat and light, that the atoms of different gases have different rates of undulation. So that increased multiformity in the arrangement of matter, has necessarily generated increased multiformity of rhythm; both through increased variety in the sizes and forms of aggregates, and through increased variety in their relations to the forces which move them. The advancing heterogeneity of motion, thus entailed by advancing heterogeneity in the distribution of matter, does not, however, end here. Besides multiplication in the kinds of rhythm, there is a progressing complexity in their combinations. As there arise wholes composed of heterogeneous parts, each of which has its own rhythm, there must arise compound rhythms proportionately heterogeneous. We before saw that this is visible even in the cyclical perturbations of the Solar System—simple as are its structure and movements. And when we contemplate highly-developed organic bodies, we find the complication of rhythms so great, that it defies definite analysis, and from moment to moment works out in resultants that are incalculable.

This conception of Evolution forms a needful complement to that on which we have hitherto chiefly dwelt. To comprehend the phenomena in their entirety, we have to contemplate both the increasing multiformity of parts, and the increasing multiformity of the actions simultaneously assumed by these parts. At the same time that there are differentiations and integrations of the matter, there are differentiations and integrations of its motion. And this increasingly heterogeneous distribution of motion, constitutes Evolution *functionally* considered; as distinguished from that increasingly heterogeneous distribution of matter, which constitutes Evolution *structurally* considered. While of course, Dissolution exhibits the transition to a reverse distribution, both structurally and functionally.

§ 107. One other preliminary must be set down. When specifically interpreting Evolution, we shall have to consider under their concrete forms, the various resolutions of force that follow its conflict with matter. Here it will be well to contemplate such resolutions under their most general or abstract forms.

Any incident force is primarily resolvable or divisible into its *effective* and *non-effective* portions. In mechanical impact, the entire momentum of a striking body is never communicated to the body struck: even under those most favourable conditions in which the striking body loses all its sensible motion, there still remains with it a portion of the original momentum, under the shape of that insensible motion produced among its particles by the collision. Of the light or heat falling on any mass, a part, more or less considerable, is reflected; and only the remaining part works molecular changes in the mass. Next it is to be noted that the effective force, is itself divisible into the *temporarily effective* and the *permanently effective*. The units of an aggregate acted on, may undergo those rhythmical changes of relative position which constitute increased vibration, as well as other changes of relative position which are not from instant to instant neutralized by opposite ones. Of these, the first, disappearing in the shape of radiating undulations, leave the molecular arrangement as it originally was; while the second conduce to that re-arrangement constituting Evolution. Yet a further distinction has to be made. The permanently effective force works out changes of relative position of two kinds—the *insensible* and the *sensible*. The insensible transpositions among the units are those constituting what we call chemical composition and decomposition; and it is these which we recognize as the qualitative differences that arise in an aggregate. The sensible transpositions are such as result when certain of the units, instead of being put into different relations with their immediate neighbours, are carried away from them and united together elsewhere.

Concerning these divisions and sub-divisions of any force affecting an aggregate, the fact which it chiefly concerns us to observe, is, that they are complementary to each other. Of the whole incident force, the effective must be that which remains after deducting the non-effective. The two parts of the effective force must vary inversely as each other: where much of it is temporarily effective, little of it can be permanently effective; and *vice versâ*. Lastly, the permanently effective force, being expended in working both the insensible re-arrangements which constitute chemical modification, and the sensible re-arrangements which result in structure, must generate of either kind an amount that is great or small in proportion as it has generated a small or great amount of the other.

§ 108. And now of the propositions grouped together in this chapter, it may be well to remark that, in common with foregoing propositions, they have for their warrant the fundamental truth with which our synthesis set out.

That when a given force falls on any aggregate, the permanently effective part of it will produce an amount of re-arrangement that is inversely proportional to the cohesion existing among the parts of the aggregate, is demonstrable *à priori*. Whether the cohesion be mechanical or chemical, or whether it be temporarily modified by a changed degree of molecular vibration, matters not to the general conclusion. In all these cases it follows from the persistence of force, that in proportion as the units offer great resistance to alteration in their relative positions, must the amount of motion which a given force impresses on them be small. The proposition is in fact an identical one; since the cohesion of units is known to be great or small, only by the smallness or greatness of the re-arrangement which a given incident force produces.

The continuity of motion we found to be a corollary from the persistence of force; and from the continuity of motion, it follows that molecular motion and the motion of masses can be respectively increased only at each other's expense. Hence, if in the course of Evolution there arises a motion of masses that did not before exist, there must have ceased an equivalent molecular motion; and if in the course of Dissolution there arises a molecular motion that did not before exist, an equivalent motion of masses must have disappeared.

Equally necessary is the conclusion that the several results of the force expended on any aggregate, must be complementary to each other. It is not less obviously a corollary from the persistence of force, that of the whole incident force the effective is the part which remains after deducting the non-effective; than it is, that of the effective force, whatever does not work permanent results, works temporary results, and that such amount of the permanently effective force as is not absorbed in producing insensible re-arrangements, will produce sensible re-arrangements.

CHAPTER XIII
THE INSTABILITY OF THE HOMOGENEOUS[16]

§ 109. Thus far our steps towards the interpretation of Evolution have been preparatory. We have dealt with the factors of the process, rather than the process itself. After the ultimate truth that, Matter, Motion, and Force, as cognizable by human intelligence, can neither come into existence nor cease to exist, we have considered certain other ultimate truths concerning the modes in which Force and Motion are manifested during the changes they produce in Matter. Now we have to study the changes themselves. We have here to analyze that re-arrangement in the parts of Matter, which occurs under the influence of Force, that is unchangeable in quantity though changeable in form, through the medium of Motion taking place rhythmically along lines of least resistance. The proposition which comes first in logical order, is, that some re-arrangement must result; and this proposition may be best dealt with under the more specific shape, that the condition of homogeneity is a condition of unstable equilibrium.

First, as to the meaning of the terms; respecting which some readers may need explanation. The phrase *unstable equilibrium* is one used in mechanics to express a balance of forces of such kind, that the interference of any further force, however minute, will destroy the arrangement previously subsisting; and bring about a totally different arrangement. Thus, a stick poised on its lower end is in unstable equilibrium: however exactly it may be placed in a perpendicular position, as soon as it is left to itself it begins, at first imperceptibly, to lean on one side, and with increasing rapidity falls into another attitude. Conversely, a stick suspended from its upper end is in stable equilibrium: however much disturbed, it will return to the same position. The proposition is, then, that the state of homogeneity, like the state of the stick poised on its lower end, is one that cannot be maintained. Let us take a few illustrations.

Of mechanical ones the most familiar is that of the scales. If they be accurately made, and not clogged by dirt or rust, it is impossible to keep a pair of scales perfectly balanced: eventually one scale will descend and the other ascend—they will assume a heterogeneous relation. Again, if we sprinkle over the surface of a fluid a number of equal-sized particles, having an attraction for each other, they will, no matter how uniformly distributed, by and by concentrate irregularly into one or more groups. Were it possible to bring a mass of water into a state of perfect homogeneity—a state of complete quiescence, and exactly equal density throughout—yet the radiation of heat from neighbouring bodies, by affecting differently its different parts, would inevitably produce inequalities of density and consequent currents; and would so render it to that extent heterogeneous. Take a piece of red-hot matter, and however evenly heated it may at first be, it will quickly cease to be so: the exterior, cooling faster than the interior, will become different in temperature from it. And the lapse into heterogeneity of temperature, so obvious in this extreme case, takes place more or less in all cases. The action of chemical forces supplies other illustrations. Expose a fragment of metal to air or water, and in course of time it will be coated with a film of oxide, carbonate, or other compound: that is—its outer parts will become unlike its inner parts. Usually the heterogeneity produced by the action of chemical forces on the surfaces of masses, is not striking; because the changed portions are soon washed away, or otherwise removed. But if this is prevented, comparatively complex structures result. Quarries of trap-rock contain some striking examples. Not unfrequently a piece of trap may be found reduced, by the action of the weather, to a number of loosely-adherent coats, like those of an onion. Where the block has been quite undisturbed, we may trace the whole series of these, from the angular, irregular outer one, through successively included ones in which the shape becomes gradually rounded, ending finally in a spherical nucleus. On comparing the original mass of stone with this group of concentric coats, each of which differs from the rest in form, and probably in the state of decomposition at which it has arrived, we get a marked illustration of the multiformity to which, in lapse of time, a uniform body may be brought by external chemical action. The instability of the homogeneous is equally seen in the changes set up throughout the interior of a mass, when it consists of units that are not rigidly bound together. The atoms of a precipitate never remain separate, and equably distributed through the fluid in which they make their appearance. They aggregate either into crystalline grains, each containing an immense number of atoms, or they aggregate into flocculi, each containing a yet larger number; and where the mass of fluid is great, and the process prolonged, these flocculi do not continue equidistant, but break up into groups. That is to say, there is a destruction of the balance at first subsisting among the diffused particles, and also of the balance at first subsisting among the groups into which these particles unite. Certain solutions of non-crystalline substances in highly volatile liquids, exhibit in the course of half an hour a whole series of changes that are set up in the alleged way. If for example a little shell-lac-varnish (made by dissolving shell-lac in coal-naphtha until it is of the consistence of cream) be poured on a piece of paper, the surface of the varnish will shortly become marked by polygonal divisions, which, first appearing round the edge of the mass, spread towards its centre. Under a lense these irregular polygons of five or more sides, are seen to be severally bounded by dark lines, on each side of which there are light-coloured borders. By the addition of matter to their inner edges, the borders slowly broaden, and thus encroach on the areas of the polygons; until at length there remains nothing but a dark spot in the centre of each. At the same time the boundaries of the polygons become curved; and they end by appearing like spherical sacs pressed together; strangely simulating (but only simulating) a group of nucleated cells. Here a rapid loss of homogeneity is exhibited in three ways:—First, in the formation of the film, which is the seat of these changes; second, in the formation of the polygonal sections into which this film divides; and third, in the contrast that arises between the polygonal sections round the edge, where they are small and early formed, and those in the centre which are larger and formed later.

The instability thus variously illustrated is obviously consequent on the fact, that the several parts of any homogeneous aggregation are necessarily exposed to different forces—forces that differ either in kind or amount; and being exposed to different forces they are of necessity differently modified. The relations of outside and inside, and of comparative nearness to neighbouring sources of influence, imply the reception of influences that are unlike in quantity or quality, or both; and it follows that unlike changes will be produced in the parts thus dissimilarly acted upon.

For like reasons it is manifest that the process must repeat itself in each of the subordinate groups of units that are differentiated by the modifying forces. Each of these subordinate groups, like the original group, must gradually, in obedience to the influences acting upon it, lose its balance of parts—must pass from a uniform into a multiform state. And so on continuously. Whence indeed it is clear that not only must the homogeneous lapse into the non-homogeneous, but that the more homogeneous must tend ever to become less homogeneous. If any given whole, instead of being absolutely uniform throughout, consist of parts distinguishable from each other—if each of these parts, while somewhat unlike other parts, is uniform within itself; then, each of them being in unstable equilibrium, it follows that while the changes set up within it must render it multiform, they must at the same time render the whole more multiform than before. The general principle, now to be followed out in its applications, is thus somewhat more comprehensive than the title of the chapter implies. No demurrer to the conclusions drawn, can be based on the ground that perfect homogeneity nowhere exists; since, whether that state with which we commence be or be not one of perfect homogeneity, the process must equally be towards a relative heterogeneity.

§ 110. The stars are distributed with a three-fold irregularity. There is first the marked contrast between the plane of the milky way and other parts of the heavens, in respect of the quantities of stars within given visual areas. There are secondary contrasts of like kind in the milky way itself, which has its thick and thin places; as well as throughout the celestial spaces in general, which are much more closely strewn in some regions than in others. And there is a third order of contrasts produced by the aggregation of stars into small clusters. Besides this heterogeneity of distribution of the stars in general, considered without distinction of kinds, a further such heterogeneity is disclosed when they are classified by their differences of colour, which doubtless answer to differences of physical constitution. While the yellow stars are found in all parts of the heavens, the red and blue stars are not so: there are wide regions in which both red and blue stars are rare; there are regions in which the blue occur in considerable numbers, and there are other regions in which the red are comparatively abundant. Yet one more irregularity of like significance is presented by the nebulæ,—aggregations of matter which, whatever be their nature, most certainly belong to our sidereal system. For the nebulæ are not dispersed with anything like uniformity; but are abundant around the poles of the galactic circle and rare in the neighbourhood of its plane. No one will expect that anything like a definite interpretation of this structure can be given on the hypothesis of Evolution, or any other hypothesis. The most that can be looked for is some reason for thinking that irregularities, not improbably of these kinds, would occur in the course of Evolution, supposing it to have taken place. Any one called on to assign such reason might argue, that if the matter of which stars and all other celestial bodies consist, be assumed to have originally existed in a diffused form throughout a space far more vast even than that which our sidereal system now occupies, the instability of the homogeneous would negative its continuance in that state. In default of an absolute balance among the forces with which the dispersed particles acted on each other (which could not exist in any aggregation having limits) he might show that motion and consequent changes of distribution would necessarily result. The next step in the argument would be that in matter of such extreme tenuity and feeble cohesion there would be motion towards local centres of gravity, as well as towards the general centre of gravity; just as, to use a humble illustration, the particles of a precipitate aggregate into flocculi at the same time that they sink towards the earth. He might urge that in the one case as in the other, these smallest and earliest local aggregations must gradually divide into groups, each concentrating to its own centre of gravity,—a process which must repeat itself on a larger and larger scale. In conformity with the law that motion once set up in any direction becomes itself a cause of subsequent motion in that direction, he might further infer that the heterogeneities thus set up would tend ever to become more pronounced. Established mechanical principles would justify him in the conclusion that the motions of these irregular masses of slightly aggregated nebular matter towards their common centre of gravity must be severally rendered curvelinear, by the resistance of the medium from which they were precipitated; and that in consequence of the irregularities of distribution already set up, such conflicting curvelinear motions must, by composition of forces, end in a rotation of the incipient sidereal system. He might without difficulty show that the resulting centrifugal force must so far modify the process of general aggregation, as to prevent anything like uniform distribution of the stars eventually formed—that there must arise a contrast such as we see between the galactic circle and the rest of the heavens. He might draw the further not unwarrantable inference, that differences in the process of local concentration would probably result from the unlikeness between the physical conditions existing around the general axis of rotation and those existing elsewhere. To which he might add, that after the formation of distinct stars, the ever-increasing irregularities of distribution due to continuance of the same causes would produce that patchiness which distinguishes the heavens in both its larger and smaller areas. We need not here however commit ourselves to such far-reaching speculations. For the purposes of the general argument it is needful only to show, that any finite mass of diffused matter, even though vast enough to form our whole sidereal system, could not be in stable equilibrium; that in default of absolute sphericity, absolute uniformity of composition, and absolute symmetry of relation to all forces external to it; its concentration must go on with an ever-increasing irregularity; and that thus the present aspect of the heavens is not, so far as we can judge, incongruous with the hypothesis of a general evolution consequent on the instability of the homogeneous.

Descending to that more limited form of the nebular hypothesis which regards the solar system as having resulted by gradual concentration; and assuming this concentration to have advanced so far as to produce a rotating spheroid of nebulous matter; let us consider what further consequence the instability of the homogeneous necessitates. Having become oblate in figure, unlike in the densities of its centre and surface, unlike in their temperatures, and unlike in the velocities with which its parts move round their common axis, such a mass can no longer be called homogeneous; and therefore any further changes exhibited by it as a whole, can illustrate the general law, only as being changes from a more homogeneous to a less homogeneous state. Changes of this kind are to be found in the transformations of such of its parts as are still homogeneous within themselves. If we accept the conclusion of Laplace, that the equatorial portion of this rotating and contracting spheroid will at successive stages acquire a centrifugal force great enough to prevent any nearer approach to the centre round which it rotates, and will so be left behind by the inner parts of the spheroid in its still-continued contraction; we shall find, in the fate of the detached ring, a fresh exemplification of the principle we are following out. Consisting of gaseous matter, such a ring, even if absolutely uniform at the time of its detachment, cannot continue so. To maintain its equilibrium there must be an almost perfect uniformity in the action of all external forces upon it (almost, we must say, because the cohesion, even of extremely attenuated matter, might suffice to neutralize very minute disturbances); and against this the probabilities are immense. In the absence of equality among the forces, internal and external, acting on such a ring, there must be a point or points at which the cohesion of its parts is less than elsewhere—a point or points at which rupture will therefore take place. Laplace assumed that the ring would rupture at one place only; and would then collapse on itself. But this is a more than questionable assumption—such at least I know to be the opinion of an authority second to none among those now living. So vast a ring, consisting of matter having such feeble cohesion, must break up into many parts. Nevertheless, it is still inferrable from the instability of the homogeneous, that the ultimate result which Laplace predicted would take place. For even supposing the masses of nebulous matter into which such a ring separated, were so equal in their sizes and distances as to attract each other with exactly equal forces (which is infinitely improbable); yet the unequal action of external disturbing forces would inevitably destroy their equilibrium—there would be one or more points at which adjacent masses would begin to part company. Separation once commenced, would with ever-accelerating speed lead to a grouping of the masses. And obviously a like result would eventually take place with the groups thus formed; until they at length aggregated into a single mass.

Leaving the region of speculative astronomy, let us consider the Solar System as it at present exists. And here it will be well, in the first place, to note a fact which may be thought at variance with the foregoing argument— namely, the still-continued existence of Saturn's rings; and especially of the internal nebulous ring lately discovered. To the objection that the outer rings maintain their equilibrium, the reply is that the comparatively great cohesion of liquid or solid substance would suffice to prevent any slight tendency to rupture from taking effect. And that a nebulous ring here still preserves its continuity, does not really negative the foregoing conclusion; since it happens under the quite exceptional influence of those symmetrically disposed forces which the external rings exercise on it. Here indeed it deserves to be noted, that though at first sight the Saturnian system appears at variance with the doctrine that a state of homogeneity is one of unstable equilibrium, it does in reality furnish a curious confirmation of this doctrine. For Saturn is not quite concentric with his rings; and it has been proved mathematically that were he and his rings concentrically situated, they could not remain so: the homogeneous relation being unstable, would gravitate into a heterogeneous one. And this fact serves to remind us of the allied one presented throughout the whole Solar System. All orbits, whether of planets or satellites, are more or less excentric—none of them are perfect circles; and were they perfect circles they would soon become ellipses. Mutual perturbations would inevitably generate excentricities. That is to say, the homogeneous relations would lapse into heterogeneous ones.

§ 111. Already so many references have been made to the gradual formation of a crust over the originally incandescent Earth, that it may be thought superfluous again to name it. It has not, however, been before considered in connexion with the general principle under discussion. Here then it must be noted as a necessary consequence of the instability of the homogeneous. In this cooling down and solidification of the Earth's surface, we have one of the simplest, as well as one of the most important, instances, of that change from a uniform to a multiform state which occurs in any mass through exposure of its different parts to different conditions. To the differentiation of the Earth's exterior from its interior thus brought about, we must add one of the most conspicuous differentiations which the exterior itself afterwards undergoes, as being similarly brought about. Were the conditions to which the surface of the Earth is exposed, alike in all directions, there would be no obvious reason why certain of its parts should become permanently unlike the rest. But being unequally exposed to the chief external centre of force—the Sun—its main divisions become unequally modified: as the crust thickens and cools, there arises that contrast, now so decided, between the polar and equatorial regions.

Along with these most marked physical differentiations of the Earth, which are manifestly consequent on the instability of the homogeneous, there have been going on numerous chemical differentiations, admitting of similar interpretation. Without raising the question whether, as some think, the so-called simple substances are themselves compounded of unknown elements (elements which we cannot separate by artificial heat, but which existed separately when the heat of the Earth was greater than any which we can produce),—without raising this question, it will suffice the present purpose to show how, in place of that comparative homogeneity of the Earth's crust, chemically considered, which must have existed when its temperature was high, there has arisen, during its cooling, an increasing chemical heterogeneity: each element or compound, being unable to maintain its homogeneity in presence of various surrounding affinities, having fallen into heterogeneous combinations. Let us contemplate this change somewhat in detail. There is every reason to believe that at an extreme heat, the bodies we call elements cannot combine. Even under such heat as can be generated artificially, some very strong affinities yield; and the great majority of chemical compounds are decomposed at much lower temperatures. Whence it seems not improbable that, when the Earth was in its first state of incandescence, there were no chemical combinations at all. But without drawing this inference, let us set out with the unquestionable fact that the compounds which can exist at the highest temperatures, and which must therefore have been the first formed as the Earth cooled, are those of the simplest constitutions. The protoxides—including under that head the alkalies, earths, &c.—are, as a class, the most fixed compounds known: the majority of them resisting decomposition by any heat we can generate. These, consisting severally of one atom of each component element, are combinations of the simplest order—are but one degree less homogeneous than the elements themselves. More heterogeneous than these, more decomposable by heat, and therefore later in the Earth's history, are the deutoxides, tritoxides, peroxides, &c.; in which two, three, four, or more atoms of oxygen are united with one atom of metal or other base. Still less able to resist heat, are the salts; which present us with compound atoms each made up of five, six, seven, eight, ten, twelve, or more atoms, of three, if not more, kinds. Then there are the hydrated salts, of a yet greater heterogeneity, which undergo partial decomposition at much lower temperatures. After them come the further-complicated supersalts and double salts, having a stability again decreased; and so throughout. After making a few unimportant qualifications demanded by peculiar affinities, I believe no chemist will deny it to be a general law of these inorganic combinations that, other things equal, the stability decreases as the complexity increases. And then when we pass to the compounds that make up organic bodies, we find this general law still further exemplified: we find much greater complexity and much less stability. An atom of albumen, for instance, consists of 482 ultimate atoms of five different kinds. Fibrine, still more intricate in constitution, contains in each atom, 298 atoms of carbon, 49 of nitrogen, 2 of sulphur, 228 of hydrogen, and 92 of oxygen—in all, 660 atoms; or, more strictly speaking—equivalents. And these two substances are so unstable as to decompose at quite moderate temperatures; as that to which the outside of a joint of roast meat is exposed. Possibly it will be objected that some inorganic compounds, as phosphuretted hydrogen and chloride of nitrogen, are more decomposable than most organic compounds. This is true. But the admission may be made without damage to the argument. The proposition is not that *all* simple combinations are more fixed than *all* complex ones. To establish our inference it is necessary only to show that, as an *average fact*, the simple combinations can exist at a higher temperature than the complex ones. And this is wholly beyond question. Thus it is manifest that the present chemical heterogeneity of the Earth's surface has arisen by degrees as the decrease of heat has permitted; and that it has shown itself in three forms—first, in the multiplication of chemical compounds; second, in the greater number of different elements contained in the more modern of these compounds; and third, in the higher and more varied multiples in which these more numerous elements combine.

Without specifying them, it will suffice just to name the meteorologic processes eventually set up in the Earth's atmosphere, as further illustrating the alleged law. They equally display that destruction of a homogeneous state which results from unequal exposure to incident forces.

§ 112. Take a mass of unorganized but organizable matter—either the body of one of the lowest living forms, or the germ of one of the higher. Consider its circumstances. Either it is immersed in water or air, or it is contained within a parent organism. Wherever placed, however, its outer and inner parts stand differently related to surrounding agencies—nutriment, oxygen, and the various stimuli. But this is not all. Whether it lies quiescent at the bottom of the water or on the leaf of a plant; whether it moves through the water preserving some definite attitude; or whether it is in the inside of an adult; it equally results that certain parts of its surface are more exposed to surrounding agencies than other parts—in some cases more exposed to light, heat, or oxygen, and in others to the maternal tissues and their contents. Hence must follow the destruction of its original equilibrium. This may take place in one of two ways. Either the disturbing forces may be such as to overbalance the affinities of the organic elements, in which case there result those changes which are known as decomposition; or, as is ordinarily the case, such changes are induced as do not destroy the organic compounds, but only modify them: the parts most exposed to the modifying forces being most modified. To elucidate this, suppose we take a few cases.

Note first what appear to be exceptions. Certain minute animal forms present us either with no appreciable differentiations or with differentiations so obscure as to be made out with great difficulty. In the Rhizopods, the substance of the jelly-like body remains throughout life unorganized, even to the extent of having no limiting membrane; as is proved by the fact that the thread-like processes protruded by the mass, coalesce on touching each other. Whether or not the nearly allied *Amœba*, of which the less numerous and more bulky processes do not coalesce, has, as lately alleged, something like a cell-wall and a nucleus, it is clear that the distinction of parts is very slight; since particles of food pass bodily into the inside through any part of the periphery, and since when the creature is crushed to pieces, each piece behaves as the whole did. Now these cases, in which there is either no contrast of structure between exterior and interior or very little, though seemingly opposed to the above inference, are really very significant evidences of its truth. For what is the peculiarity of this division of the *Protozoa*? Its members undergo perpetual and irregular changes of form—they show no persistent relation of parts. What lately formed a portion of the interior is now protruded, and, as a temporary limb, is attached to some object it happens to touch. What is now a part of the surface will presently be drawn, along with the atom of nutriment sticking to it, into the centre of the mass. Either the relations of inner and outer have no permanent existence, or they are very slightly marked. But by the hypothesis, it is only because of their unlike positions with respect to modifying forces, that the originally like units of a living mass become unlike. We must therefore expect no established differentiation of parts in creatures which exhibit no established differences of position in their parts; and we must expect extremely little differentiation of parts where the differences of position are but little determined—which is just what we find. This negative evidence is borne out by positive evidence. When we turn from these proteiform specks of living jelly to organisms having an unchanging distribution of substance, we find differences of tissue corresponding to differences of relative position. In all the higher *Protozoa*, as also in the *Protophyta*, we meet with a fundamental differentiation into cell-membrane and cell-contents; answering to that fundamental contrast of conditions implied by the terms outside and inside. On passing from what are roughly classed as unicellular organisms, to the lowest of those which consist of aggregated cells, we equally observe the connection between structural differences and differences of circumstance. Negatively, we see that in the sponge, permeated throughout by currents of sea-water, the indefiniteness of organization corresponds with the absence of definite unlikeness of conditions: the peripheral and central portions are as little contrasted in structure as in exposure to surrounding agencies. While positively, we see that in a form like the *Thalassicolla*, which, though equally humble, maintains its outer and inner parts in permanently unlike circumstances, there is displayed a rude structure obviously subordinated to the primary relations of centre and surface: in all its many and important varieties, the parts exhibit a more or less concentric arrangement.

After this primary modification, by which the outer tissues are differentiated from the inner, the next in order of constancy and importance is that by which some part of the outer tissues is differentiated from the rest; and this corresponds with the almost universal fact that some part of the outer tissues is more exposed to certain environing influences than the rest. Here, as before, the apparent exceptions are extremely significant. Some of the lowest vegetal organisms, as the *Hematococci* and *Protococci*, evenly imbedded in a mass of mucus, or dispersed through the Arctic snow, display no differentiations of surface; the several parts of their surfaces being subjected to no definite contrasts of conditions. Ciliated spheres such as the *Volvox* have no parts of their periphery unlike other parts; and it is not to be expected that they should have; since, as they revolve in all directions, they do not, in traversing the water, permanently expose any part to special conditions. But when we come to organisms that are either fixed, or while moving preserve definite attitudes, we no longer find uniformity of surface. The most general fact which can be asserted with respect to the structures of plants and animals, is, that however much alike in shape and texture the various parts of the exterior may at first be, they acquire unlikenesses corresponding to the unlikenesses of their relations to surrounding agencies. The ciliated germ of a Zoophyte, which, during its locomotive stage, is distinguishable only into outer and inner tissues, no sooner becomes fixed, than its upper end begins to assume a different structure from its lower. The disc-shaped *gemmæ* of the *Marchantia*, originally alike on both surfaces, and falling at random with either side uppermost, immediately begin to develop rootlets on the under side, and *stomata* on the upper side: a fact proving beyond question, that this primary differentiation is determined by this fundamental contrast of conditions.

Of course in the germs of higher organisms, the metamorphoses immediately due to the instability of the homogeneous, are soon masked by those due to the assumption of the hereditary type. Such early changes, however, as are common to all classes of organisms, and so cannot be ascribed to heredity, entirely conform to the hypothesis. A germ which has undergone no developmental modifications, consists of a spheroidal group of homogeneous cells. Universally, the first step in its evolution is the establishment of a difference between some of the peripheral cells and the cells which form the interior—some of the peripheral cells, after repeated spontaneous fissions, coalesce into a membrane; and by continuance of the process this membrane spreads until it speedily invests the entire mass, as in mammals, or, as in birds, stops short of that for some time. Here we have two significant facts. The first is, that the primary unlikeness arises between the exterior and the interior. The second is, that the change which thus initiates development, does not take place simultaneously over the whole exterior; but commences at one place, and gradually involves the rest. Now these facts are just those which might be inferred from the instability of the homogeneous. The surface must, more than any other part, become unlike the centre, because it is most dissimilarly conditioned; and all parts of the surface cannot simultaneously exhibit this differentiation, because they cannot be exposed to the incident forces with absolute uniformity. One other general fact of like implication remains. Whatever be the extent of this peripheral layer of cells, or blastoderm as it is called, it presently divides into two layers—the serous and mucous; or, as they have been otherwise called, the ectoderm and the endoderm. The first of these is formed from that portion of the layer which lies in contact with surrounding agents; and the second of them is formed from that portion of the layer which lies in contact with the contained mass of yelk. That is to say, after the primary differentiation, more or less extensive, of surface from centre, the resulting superficial portion undergoes a secondary differentiation into inner and outer parts—a differentiation which is clearly of the same order with the preceding, and answers to the next most marked contrast of conditions.

But, as already hinted, this principle, understood in the simple form here presented, supplies no key to the detailed phenomena of organic development. It fails entirely to explain generic and specific peculiarities; and indeed leaves us equally in the dark respecting those more important distinctions by which families and orders are marked out. Why two ova, similarly exposed in the same pool, should become the one a fish, and the other a reptile, it cannot tell us. That from two different eggs placed under the same hen, should respectively come forth a duckling and a chicken, is a fact not to be accounted for on the hypothesis above developed. We have here no alternative but to fall back upon the unexplained principle of hereditary transmission. The capacity possessed by an unorganized germ of unfolding into a complex adult, which repeats ancestral traits in the minutest details, and that even when it has been placed in conditions unlike those of its ancestors, is a capacity we cannot at present understand. That a microscopic portion of seemingly structureless matter should embody an influence of such kind, that the resulting man will in fifty years after become gouty or insane, is a truth which would be incredible were it not daily illustrated. Should it however turn out, as we shall hereafter find reason for suspecting, that these complex differentiations which adults exhibit, are themselves the slowly accumulated and transmitted results of a process like that seen in the first changes of the germ; it will follow that even those embryonic changes due to hereditary influence, are remote consequences of the alleged law. Should it be shown that the slight modifications wrought during life on each adult, and bequeathed to offspring along with all like preceding modifications, are themselves unlikenesses of parts that are produced by unlikenesses of conditions; then it will follow that the modifications displayed in the course of embryonic development, are partly direct consequences of the instability of the homogeneous, and partly indirect consequences of it. To give reasons for entertaining this hypothesis, however, is not needful for the justification of the position here taken. It is enough that the most conspicuous differentiations which incipient organisms universally display, correspond to the most marked differences of conditions to which their parts are subject. It is enough that the habitual contrast between outside and inside, which we *know* is produced in inorganic masses by unlikeness of exposure to incident forces, is strictly paralleled by the first contrast that makes its appearance in all organic masses.

It remains to point out that in the assemblage of organisms constituting a species, the principle enunciated is equally traceable. We have abundant materials for the induction that each species will not remain uniform, but is ever becoming to some extent multiform; and there is ground for the deduction that this lapse from homogeneity to heterogeneity is caused by the subjection of its members to unlike sets of circumstances. The fact that in every species, animal and vegetal, the individuals are never quite alike; joined with the fact that there is in every species a tendency to the production of differences marked enough to constitute varieties; form a sufficiently wide basis for the induction. While the deduction is confirmed by the familiar experience that varieties are most numerous and decided where, as among cultivated plants and domestic animals, the conditions of life depart from the original ones, most widely and in the most numerous ways. Whether we regard "natural selection" as wholly, or only in part, the agency through which varieties are established, matters not to the general conclusion. For as the survival of any variety proves its constitution to be in harmony with a certain aggregate of surrounding forces—as the multiplication of a variety and the usurpation by it of an area previously occupied by some other part of the species, implies different effects produced by such aggregate of forces on the two, it is clear that this aggregate of forces is the real cause of the differentiation—it is clear that if the variety supplants the original species in some localities but not in others, it does so because the aggregate of forces in the one locality is unlike that in the other—it is clear that the lapse of the species from a state of homogeneity to a state of heterogeneity arises from the exposure of its different parts to different aggregates of forces.

§ 113. Among mental phenomena it is difficult to establish the alleged law without an analysis too extensive for the occasion. To show satisfactorily how states of consciousness, originally homogeneous, become heterogeneous through differences in the changes wrought by different forces, would require us carefully to trace out the organization of early experiences. Were this done it would become manifest that the development of intelligence, is, under one of its chief aspects, a dividing into separate classes, the unlike things previously confounded together in one class—a formation of sub-classes and sub-sub-classes, until the once confused aggregate of objects known, is resolved into an aggregate which unites extreme heterogeneity among its multiplied groups, with complete homogeneity among the members of each group. If, for example, we followed, through ascending grades of creatures, the genesis of that vast structure of knowledge acquired by sight, we should find that in the first stage, where eyes suffice for nothing beyond the discrimination of light from darkness, the only possible classifications of objects seen, must be those based on the manner in which light is obstructed, and the degree in which it is obstructed. We should find that by such undeveloped visual organs, the shadows traversing the rudimentary retina would be merely distinguished into those of the stationary objects which the creature passed during its own movements, and those of the moving objects which came near the creature while it was at rest; and that so the extremely general classification of visible things into stationary and moving, would be the earliest formed. We should find that whereas the simplest eyes are not fitted to distinguish between an obstruction of light caused by a small object close to, and an obstruction caused by a large object at some distance, eyes a little more developed must be competent to such a distinction; whence must result a vague differentiation of the class of moving objects, into the nearer and the more remote. We should find that such further improvements in vision as those which make possible a better estimation of distances by adjustment of the optic axes, and those which, through enlargement and subdivision of the retina, make possible the discrimination of shapes, must have the effects of giving greater definiteness to the classes already formed, and of sub-dividing these into smaller classes, consisting of objects less unlike. And we should find that each additional refinement of the perceptive organs, must similarly lead to a multiplication of divisions and a sharpening of the limits of each division. In every infant might be traced the analogous transformation of a confused aggregate of impressions of surrounding objects, not recognized as differing in their distances, sizes, and shapes, into separate classes of objects unlike each other in these and various other respects. And in the one case as in the other, it might be shown that the change from this first indefinite, incoherent and comparatively homogeneous consciousness, to a definite, coherent, and heterogeneous one, is due to differences in the actions of incident forces on the organism. These brief indications of what might be shown, did space permit, must here suffice. Probably they will give adequate clue to an argument by which each reader may satisfy himself that the course of mental evolution offers no exception to the general law. In further aid of such an argument, I will here add an illustration that is comprehensible apart from the process of mental evolution as a whole.

It has been remarked (I am told by Coleridge, though I have been unable to find the passage) that with the advance of language, words which were originally alike in their meanings acquire unlike meanings—a change which he expresses by the formidable word "desynonymization." Among indigenous words this loss of equivalence cannot be clearly shown; because in them the divergencies of meaning began before the dawn of literature. But among words that have been coined, or adopted from other languages, since the writing of books commenced, it is demonstrable. In the old divines, *miscreant* is used in its etymological sense of *unbeliever*; but in modern speech it has entirely lost this sense. Similarly with *evil-doer* and *malefactor*: exactly synonymous as these are by derivation, they are no longer synonymous by usage: by a *malefactor* we now understand a convicted criminal, which is far from being the acceptation of *evil-doer*. The verb *produce*, bears in Euclid its primary meaning—to *prolong*, or *draw out*; but the now largely developed meanings of *produce* have little in common with the meanings of *prolong*, or *draw out*. In the Church of England liturgy, an odd effect results from the occurrence of *prevent* in its original sense— *to come before*, instead of its modern specialized sense—*to come before with the effect of arresting*. But the most conclusive cases are those in which the contrasted words consist of the same parts differently combined; as in *go under* and *undergo*. We *go under* a tree, and we *undergo* a pain. But though, if analytically considered, the meanings of these expressions would be the same were the words transposed, habit has so far modified their meanings that we could not without absurdity speak of *undergoing* a tree and *going under* a pain. Countless such instances might be brought to show that between two words which are originally of like force, an equilibrium cannot be maintained. Unless they are daily used in exactly equal degrees, in exactly similar relations (against which there are infinite probabilities), there necessarily arises a habit of associating one rather than the other with particular acts, or objects. Such a habit, once commenced, becomes confirmed; and gradually their homogeneity of meaning disappears. In each individual we may see the tendency which inevitably leads to this result. A certain vocabulary and a certain set of phrases, distinguish the speech of each person: each person habitually uses certain words in places where other words are habitually used by other persons; and there is a continual recurrence of favourite expressions. This inability to maintain a balance in the use of verbal symbols, which characterizes every man, characterizes, by consequence, aggregates of men; and the desynonymization of words is the ultimate effect.

Should any difficulty be felt in understanding how these mental changes exemplify a law of physical transformations that are wrought by physical forces, it will disappear on contemplating acts of mind as nervous functions. It will be seen that each loss of equilibrium above instanced, is a loss of functional equality between some two elements of the nervous system. And it will be seen that, as in other cases, this loss of functional equality is due to differences in the incidence of forces.

§ 114. Masses of men, in common with all other masses, show a like proclivity similarly caused. Small combinations and large societies equally manifest it; and in the one, as in the other, both governmental and industrial differentiations are initiated by it. Let us glance at the facts under these two heads.

A business partnership, balanced as the authorities of its members may theoretically be, practically becomes a union in which the authority of one partner is tacitly recognized as greater than that of the other or others. Though the shareholders have given equal powers to the directors of their company, inequalities of power soon arise among them; and usually the supremacy of some one director grows so marked, that his decisions determine the course which the board takes. Nor in associations for political, charitable, literary, or other purposes, do we fail to find a like process of division into dominant and subordinate parties; each having its leader, its members of less influence, and its mass of uninfluential members. These minor instances in which unorganized groups of men, standing in homogeneous relations, may be watched gradually passing into organized groups of men standing in heterogeneous relations, give us the key to social inequalities. Barbarous and civilized communities are alike characterized by separation into classes, as well as by separation of each class into more important and less important units; and this structure is manifestly the gradually-consolidated result of a process like that daily exemplified in trading and other combinations. So long as men are constituted to act on one another, either by physical force or by force of character, the struggles for supremacy must finally be decided in favour of some one; and the difference once commenced must tend to become ever more marked. Its unstable equilibrium being destroyed, the uniform must gravitate with increasing rapidity into the multiform. And so supremacy and subordination must establish themselves, as we see they do, throughout the whole structure of a society, from the great class-divisions pervading its entire body, down to village cliques, and even down to every posse of school-boys. Probably it will be objected that such changes result, not from the homogeneity of the original aggregations, but from their non-homogeneity—from certain slight differences existing among their units at the outset. This is doubtless the proximate cause. In strictness, such changes must be regarded as transformations of the relatively homogeneous into the relatively heterogeneous. But it is abundantly clear that an aggregation of men, absolutely alike in their endowments, would eventually undergo a similar transformation. For in the absence of perfect uniformity in the lives severally led by them—in their occupations, physical conditions, domestic relations, and trains of thought and feeling—there must arise differences among them; and these must finally initiate social differentiations. Even inequalities of health caused by accidents, must, by entailing inequalities of physical and mental power, disturb the exact balance of mutual influences among the units; and the balance once disturbed, must inevitably be lost. Whence, indeed, besides seeing that a body of men absolutely homogeneous in their governmental relations, must, like all other homogeneous bodies, become heterogeneous, we also see that it must do this from the same ultimate cause—unequal exposure of its parts to incident forces.

The first industrial divisions of societies are much more obviously due to unlikenesses of external circumstances. Such divisions are absent until such unlikenesses are established. Nomadic tribes do not permanently expose any groups of their members to special local conditions; nor does a stationary tribe, when occupying only a small area, maintain from generation to generation marked contrasts in the local conditions of its members; and in such tribes there are no decided economical differentiations. But a community which, growing populous, has overspread a large tract, and has become so far settled that its members live and die in their respective districts, keeps its several sections in different physical circumstances; and then they no longer remain alike in their occupations. Those who live dispersed continue to hunt or cultivate the earth; those who spread to the sea-shore fall into maritime occupations; while the inhabitants of some spot chosen, perhaps for its centrality, as one of periodical assemblage, become traders, and a town springs up. Each of these classes undergoes a modification of character consequent on its function, and better fitting it to its function. Later in the process of social evolution these local adaptations are greatly multiplied. A result of differences in soil and climate, is that the rural inhabitants in different parts of the kingdom have their occupations partially specialized; and become respectively distinguished as chiefly producing cattle, or sheep, or wheat, or oats, or hops, or cyder. People living where coal-fields are discovered are transformed into colliers; Cornishmen take to mining because Cornwall is metalliferous; and the iron-manufacture is the dominant industry where ironstone is plentiful. Liverpool has assumed the office of importing cotton, in consequence of its proximity to the district where cotton goods are made; and for analogous reasons, Hull has become the chief port at which foreign wools are brought in. Even in the establishment of breweries, of dye-works, of slate-quarries, of brickyards, we may see the same truth. So that both in general and in detail, the specializations of the social organism which characterize separate districts, primarily depend on local circumstances. Those divisions of labour which under another aspect were interpreted as due to the setting up of motion in the directions of least resistance (§ 91), are here interpreted as due to differences in the incident forces; and the two interpretations are quite consistent with each other. For that which in each case *determines* the direction of least resistance, is the distribution of the forces to be overcome; and hence unlikenesses of distribution in separate localities, entails unlikenesses in the course of human action in those localities—entails industrial differentiations.

§ 115. In common with the general truths set forth in preceding chapters, the instability of the homogeneous is demonstrable *à priori*. It, like each of them, is a corollary from the persistence of force. Already this has been tacitly implied by assigning unlikeness in the exposure of its part to surrounding agencies, as the reason why a uniform mass loses its uniformity. But here it will be proper to expand this tacit implication into definite proof.

On striking a mass of matter with such force as either to indent it or make it fly to pieces, we see both that the blow affects differently its different parts, and that the differences are consequent on the unlike relations of its parts to the force impressed. The part with which the striking body comes in contact, receiving the whole of the communicated momentum, is driven in towards the centre of the mass. It thus compresses and tends to displace the more centrally situated portions of the mass. These, however, cannot be compressed or thrust out of their places without pressing on all surrounding portions. And when the blow is violent enough to fracture the mass, we see, in the radial dispersion of its fragments, that the original momentum, in being distributed throughout it, has been divided into numerous minor momenta, unlike in their directions. We see that these directions are determined by the positions of the parts with respect to each other, and with respect to the point of impact. We see that the parts are differently affected by the disruptive force, because they are differently related to it in their directions and attachments—that the effects being the joint products of the cause and the conditions, cannot be alike in parts which are differently conditioned. A body on which radiant heat is falling, exemplifies this truth still more clearly. Taking the simplest case (that of a sphere) we see that while the part nearest to the radiating centre receives the rays at right angles, the rays strike the other parts of the exposed side at all angles from 90°

down to 0°. Again, the molecular vibrations propagated through the mass from the surface which receives the heat, must proceed inwards at angles differing for each point. Further, the interior parts of the sphere affected by the vibrations proceeding from all points of the heated side, must be dissimilarly affected in proportion as their positions are dissimilar. So that whether they be on the recipient area, in the middle, or at the remote side, the constituent atoms are all thrown into states of vibration more or less unlike each other.

But now, what is the ultimate meaning of the conclusion that a uniform force produces different changes throughout a uniform mass, because the parts of the mass stand in different relations to the force? Fully to understand this, we must contemplate each part as simultaneously subject to other forces—those of gravitation, of cohesion, of molecular motion, &c. The effect wrought by an additional force, must be a resultant of it and the forces already in action. If the forces already in action on two parts of any aggregate, are different in their directions, the effects produced on these two parts by like forces must be different in their directions. Why must they be different? They must be different because such unlikeness as exists between the two sets of factors, is made by the presence in the one of some specially-directed force that is not present in the other; and that this force will produce an effect, rendering the total result in the one case unlike that in the other, is a necessary corollary from the persistence of force. Still more manifest does it become that the dissimilarly-placed parts of any aggregate must be dissimilarly modified by an incident force, when we remember that the *quantities* of the incident force to which they are severally subject, are not equal, as above supposed; but are nearly always very unequal. The outer parts of masses are usually alone exposed to chemical actions; and not only are their inner parts shielded from the affinities of external elements, but such affinities are brought to bear unequally on their surfaces; since chemical action sets up currents through the medium in which it takes place, and so brings to the various parts of the surface unequal quantities of the active agent. Again, the amounts of any external radiant force which the different parts of an aggregate receive, are widely contrasted: we have the contrast between the quantity falling on the side next the radiating centre, and the quantity, or rather no quantity, falling on the opposite side; we have contrasts in the quantities received by differently-placed areas on the exposed side; and we have endless contrasts between the quantities received by the various parts of the interior. Similarly when mechanical force is expended on any aggregate, either by collision, continued pressure, or tension, the amounts of strain distributed throughout the mass are manifestly unlike for unlike positions. But to say the different parts of an aggregate receive different quantities of any incident force, is to say that their states are modified by it in different degrees—is to say that if they were before homogeneous in their relations they must be rendered to a proportionate extent heterogeneous; since, force being persistent, the different quantities of it falling on the different parts, must work in them different quantities of effect—different changes. Yet one more kindred deduction is required to complete the argument. We may, by parallel reasoning, reach the conclusion that, even apart from the action of any external force, the equilibrium of a homogeneous aggregate must be destroyed by the unequal actions of its parts on each other. That mutual influence which produces aggregation (not to mention other mutual influences) must work different effects on the different parts; since they are severally exposed to it in unlike amounts and directions. This will be clearly seen on remembering that the portions of which the whole is made up, may be severally regarded as minor wholes; that on each of these minor wholes, the action of the entire aggregate then becomes an external incident force; that such external incident force must, as above shown, work unlike changes in the parts of any such minor whole; and that if the minor wholes are severally thus rendered heterogeneous, the entire aggregate is rendered heterogeneous.

The instability of the homogeneous is thus deducible from that primordial truth which underlies our intelligence. One stable homogeneity only, is hypothetically possible. If centres of force, absolutely uniform in their powers, were diffused with absolute uniformity through unlimited space, they would remain in equilibrium. This however, though a verbally intelligible supposition, is one that cannot be represented in thought; since unlimited space is inconceivable. But all finite forms of the homogeneous— all forms of it which we can know or conceive, must inevitably lapse into heterogeneity. In three several ways does the persistence of force necessitate this. Setting external agencies aside, each unit of a homogeneous whole must be differently affected from any of the rest by the aggregate action of the rest upon it. The resultant force exercised by the aggregate on each unit, being in no two cases alike in both amount and direction, and usually not in either, any incident force, even if uniform in amount and direction, cannot produce like effects on the units. And the various positions of the parts in relation to any incident force, preventing them from receiving it in uniform amounts and directions, a further difference in the effects wrought on them is inevitably produced.

One further remark is needed. To the conclusion that the changes with which Evolution *commences*, are thus necessitated, remains to be added the conclusion that these changes must *continue*. The absolutely homogeneous must lose its equilibrium; and the relatively homogeneous must lapse into the relatively less homogeneous. That which is true of any total mass, is true of the parts into which it segregates. The uniformity of each such part must as inevitably be lost in multiformity, as was that of the original whole; and for like reasons. And thus the continued changes which characterize Evolution, in so far as they are constituted by the lapse of the homogeneous into the heterogeneous, and of the less heterogeneous into the more heterogeneous, are necessary consequences of the persistence of force.

16. The idea developed in this chapter originally formed part of an article on "Transcendental Physiology," published in 1857. See *Essays*, pp. 279–290.

CHAPTER XIV
THE MULTIPLICATION OF EFFECTS

§ 116. To the cause of increasing complexity set forth in the last chapter, we have in this chapter to add another. Though secondary in order of time, it is scarcely secondary in order of importance. Even in the absence of the cause already assigned, it would necessitate a change from the homogeneous to the heterogeneous; and joined with it, it makes this change both more rapid and more involved. To come in sight of it, we have but to pursue a step further, that conflict between force and matter already delineated. Let us do this.

When a uniform aggregate is subject to a uniform force, we have seen that its constituents, being differently conditioned, are differently modified. But while we have contemplated the various parts of the aggregate as thus undergoing unlike changes, we have not yet contemplated the unlike changes simultaneously produced on the various parts of the incident force. These must be as numerous and important as the others. Action and re-action being equal and opposite, it follows that in differentiating the parts on which it falls in unlike ways, the incident force must itself be correspondingly differentiated. Instead of being as before, a uniform force, it must thereafter be a multiform force—a group of dissimilar forces. A few illustrations will make this truth manifest.

A single force is divided by conflict with matter into forces that widely diverge. In the case lately cited, of a body shattered by violent collision, besides the change of the homogeneous mass into a heterogeneous group of scattered fragments, there is a change of the homogeneous momentum into a group of momenta, heterogeneous in both amounts and directions. Similarly with the forces we know as light and heat. After the dispersion of these by a radiating body towards all points, they are re-dispersed towards all points by the bodies on which they fall. Of the Sun's rays, issuing from him on every side, some few strike the Moon. These being reflected at all angles from the Moon's surface, some few of them strike the Earth. By a like process the few which reach the Earth are again diffused through surrounding space. And on each occasion, such portions of the rays as are absorbed instead of reflected, undergo refractions that equally destroy their parallelism. More than this is true. By conflict with matter, a uniform force is in part changed into forces differing in their directions; and in part it is changed into forces differing in their kinds. When one body is struck against another, that which we usually regard as the effect, is a change of position or motion in one or both bodies. But a moment's thought shows that this is a very incomplete view of the matter. Besides the visible mechanical result, sound is produced; or, to speak accurately, a vibration in one or both bodies, and in the surrounding air: and under some circumstances we call this the effect. Moreover, the air has not simply been made to vibrate, but has had currents raised in it by the transit of the bodies. Further, if there is not that great structural change which we call fracture, there is a disarrangement of the particles of the two bodies around their point of collision; amounting in some cases to a visible condensation. Yet more, this condensation is accompanied by disengagement of heat. In some cases a spark—that is, light—results, from the incandescence of a portion struck off; and occasionally this incandescence is associated with chemical combination. Thus, by the original mechanical force expended in the collision, at least five, and often more, different kinds of forces have been produced. Take, again, the lighting of a candle. Primarily, this is a chemical change consequent on a rise of temperature. The process of combination having once been set going by extraneous heat, there is a continued formation of carbonic acid, water, &c.—in itself a result more complex than the extraneous heat that first caused it. But along with this process of combination there is a production of heat; there is a production of light; there is an ascending column of hot gases generated; there are currents established in the surrounding air. Nor does the decomposition of one force into many forces end here. Each of the several changes worked becomes the parent of further changes. The carbonic acid formed, will by and by combine with some base; or under the influence of sunshine give up its carbon to the leaf of a plant. The water will modify the hygrometric state of the air around; or, if the current of hot gases containing it come against a cold body, will be condensed: altering the temperature, and perhaps the chemical state, of the surface it covers. The heat given out melts the subjacent tallow, and expands whatever it warms. The light, falling on various substances, calls forth from them reactions by which it is modified; and so divers colours are produced. Similarly even with these secondary actions, which may be traced out into ever-multiplying ramifications, until they become too minute to be appreciated. Universally, then, the effect is more complex than the cause. Whether the aggregate on which it falls be homogeneous or otherwise, an incident force is transformed by the conflict into a number of forces that differ in their amounts, or directions, or kinds; or in all these respects. And of this group of variously-modified forces, each ultimately undergoes a like transformation.

Let us now mark how the process of evolution is furthered by this multiplication of effects. An incident force decomposed by the reactions of a body into a group of unlike forces—a uniform force thus reduced to a multiform force—becomes the cause of a secondary increase of multiformity in the body which decomposes it. In the last chapter we saw that the several parts of an aggregate are differently modified by any incident force. It has just been shown that by the reactions of the differently modified parts, the incident force itself must be divided into differently modified parts. Here it remains to point out that each differentiated division of the aggregate, thus becomes a centre from which a differentiated division of the original force is again diffused. And since unlike forces must produce unlike results, each of these differentiated forces must produce, throughout the aggregate, a further series of differentiations. This secondary cause of the change from homogeneity to heterogeneity, obviously becomes more potent in proportion as the heterogeneity increases. When the parts into which any evolving whole has segregated itself, have diverged widely in nature, they will necessarily react very diversely on any incident force—they will divide an incident force into so many strongly contrasted groups of forces. And each of them becoming the centre of a quite distinct set of influences, must add to the number of distinct secondary changes wrought throughout the aggregate. Yet another corollary must be added. The number of unlike parts of which an aggregate consists, as well as the degree of their unlikeness, is an important factor in the process. Every additional specialized division is an additional centre of specialized forces. If a uniform whole, in being itself made multiform by an incident force, makes the incident force multiform; if a whole consisting of two unlike sections, divides an incident force into two unlike groups of multiform forces; it is clear that each new unlike section must be a further source of complication among the forces at work throughout the mass—a further source of heterogeneity. The multiplication of effects must proceed in geometrical progression. Each stage of evolution must initiate a higher stage.

§ 117. The force of aggregation acting on irregular masses of rare matter, diffused through a resisting medium, will not cause such masses to move in straight lines to their common centre of gravity; but, as before said, each will take a curvilinear path, directed to one or other side of the centre of gravity. All of them being differently conditioned, gravitation will impress on each a motion differing in direction, in velocity, and in the degree of its curvature— uniform aggregative force will be differentiated into multiform momenta. The process thus commenced, must go on till it produces a single mass of nebulous matter; and these independent curvilinear motions must result in a movement of this mass round its axis: a simultaneous condensation and rotation in which we see how two effects of the aggregative force, at first but slightly divergent, become at last widely differentiated. A gradual increase of oblateness in this revolving spheroid, must take place through the joint action of these two forces, as the bulk diminishes and the rotation grows more rapid; and this we may set down as a third effect. The genesis of heat, which must accompany augmentation of density, is a consequence of yet another order—a consequence by no means simple; since the various parts of the mass, being variously condensed, must be variously heated. Acting throughout a gaseous spheroid, of which the parts are unlike in their temperatures, the forces of aggregation and rotation must work a further series of changes: they must set up circulating currents, both general and local. At a later stage light as well as heat will be generated. Thus without dwelling on the likelihood of chemical combinations and electric disturbances, it is sufficiently manifest that, supposing matter to have originally existed in a diffused state, the once uniform force which caused its aggregation, must have become gradually divided into different forces; and that each further stage of complication in the resulting aggregate, must have initiated further subdivisions of this force—a further multiplication of effects, increasing the previous heterogeneity.

This section of the argument may however be adequately sustained, without having recourse to any such hypothetical illustrations as the foregoing. The astronomical attributes of the Earth, will even alone suffice our purpose. Consider first the effects of its momentum round its axis. There is the oblateness of its form; there is the alternation of day and night; there are certain constant marine currents; and there are certain constant aërial currents. Consider next the secondary series of consequences due to the divergence of the Earth's plane of rotation from the plane of its orbit. The many differences of the seasons, both simultaneous and successive, which pervade its surface, are thus caused. External attraction acting on this rotating oblate spheroid with inclined axis, produces the motion called nutation, and that slower and larger one from which follows the precession of the equinoxes, with its several sequences. And then by this same force are generated the tides, aqueous and atmospheric.

Perhaps, however, the simplest way of showing the multiplication of effects among phenomena of this order, will be to set down the influences of any member of the Solar System on the rest. A planet directly produces in neighbouring planets certain appreciable perturbations, complicating those otherwise produced in them; and in the remoter planets it directly produces certain less visible perturbations. Here is a first series of effects. But each of the perturbed planets is itself a source of perturbations—each directly affects all the others. Hence, planet A having drawn planet B out of the position it would have occupied in A's absence, the perturbations which B causes are different from what they would else have been; and similarly with C, D, E, &c. Here then is a secondary series of effects: far more numerous though far smaller in their amounts. As these indirect perturbations must to some extent modify the movements of each planet, there results from them a tertiary series; and so on continually. Thus the force exercised by any planet works a different effect on each of the rest; this different effect is from each as a centre partially broken up into minor different effects on the rest; and so on in ever multiplying and diminishing waves throughout the entire system.

§ 118. If the Earth was formed by the concentration of diffused matter, it must at first have been incandescent; and whether the nebular hypothesis be accepted or not, this original incandescence of the Earth must now be regarded as inductively established—or, if not established, at least rendered so probable that it is a generally admitted geological doctrine. Several results of the gradual cooling of the Earth—as the formation of a crust, the solidification of sublimed elements, the precipitation of water, &c., have been already noticed—and I here again refer to them merely to point out that they are simultaneous effects of the one cause, diminishing heat. Let us now, however, observe the multiplied changes afterwards arising from the continuance of this one cause. The Earth, falling in temperature, must contract. Hence the solid crust at any time existing, is presently too large for the shrinking nucleus; and being unable to support itself, inevitably follows the nucleus. But a spheroidal envelope cannot sink down into contact with a smaller internal spheroid, without disruption: it will run into wrinkles, as the rind of an apple does when the bulk of its interior decreases from evaporation. As the cooling progresses and the envelope thickens, the ridges consequent on these contractions must become greater; rising ultimately into hills and mountains; and the later systems of mountains thus produced must not only be higher, as we find them to be, but they must be longer, as we also find them to be. Thus, leaving out of view other modifying forces, we see what immense heterogeneity of surface arises from the one cause, loss of heat—a heterogeneity which the telescope shows us to be paralleled on the Moon, where aqueous and atmospheric agencies have been absent. But we have yet to notice another kind of heterogeneity of surface, similarly and simultaneously caused. While the Earth's crust was still thin, the ridges produced by its contraction must not only have been small, but the tracts between them must have rested with comparative smoothness on the subjacent liquid spheroid; and the water in those arctic and antarctic regions where it first condensed, must have been evenly distributed. But as fast as the crust grew thicker and gained corresponding strength, the lines of fracture from time to time caused in it, necessarily occurred at greater distances apart; the intermediate surfaces followed the contracting nucleus with less uniformity; and there consequently resulted larger areas of land and water. If any one, after wrapping an orange in wet tissue paper, and observing both how small are the wrinkles and how evenly the intervening spaces lie on the surface of the orange, will then wrap it in thick cartridge-paper, and note both the greater height of the ridges and the larger spaces throughout which the paper does not touch the orange, he will realize the fact, that as the Earth's solid envelope thickened, the areas of elevation and depression became greater. In place of islands more or less homogeneously scattered over an all-embracing sea, there must have gradually arisen heterogeneous arrangements of continent and ocean, such as we now know. This double change in the extent and in the elevation of the lands, involved yet another species of heterogeneity—that of coast-line. A tolerably even surface raised out of the ocean will have a simple, regular sea-margin; but a surface varied by table-lands and intersected by mountain-chains, will, when raised out of the ocean, have an outline extremely irregular, alike in its leading features and in its details. Thus endless is the accumulation of geological and geographical results slowly brought about by this one cause—the escape of the Earth's primitive heat.

When we pass from the agency which geologists term igneous, to aqueous and atmospheric agencies, we see a like ever-growing complication of effects. The denuding actions of air and water have, from the beginning, been modifying every exposed surface: everywhere working many different changes. As already shown (§ 80) the original source of those gaseous and fluid motions which effect denudation, is the solar heat. The transformation of this into various modes of force, according to the nature and condition of the matter on which it falls, is the first stage of complication. The sun's rays, striking at all angles a sphere, that from moment to moment presents and withdraws different parts of its surface, and each of them for a different time daily throughout the year, would produce a considerable variety of changes even were the sphere uniform. But falling as they do on a sphere surrounded by an atmosphere in some parts of which wide areas of cloud are suspended, and which here unveils vast tracts of sea, there of level land, there of mountains, there of snow and ice, they initiate in its several parts countless different movements. Currents of air of all sizes, directions, velocities, and temperatures, are set up; as are also marine currents similarly contrasted in their characters. In this region the surface is giving off water in the state of vapour; in that, dew is being precipitated; and in the other rain is descending—differences that arise from the ever-changing ratio between the absorption and radiation of heat in each place. At one hour, a rapid fall in temperature leads to the formation of ice, with an accompanying expansion throughout the moist bodies frozen; while at another, a thaw unlocks the dislocated fragments of these bodies. And then, passing to a second stage of complication, we see that the many kinds of motion directly or indirectly caused by the sun's rays, severally produce results that vary with the conditions. Oxidation, drought, wind, frost, rain, glaciers, rivers, waves, and other denuding agents effect disintegrations that are determined in their amounts and qualities by local circumstances. Acting upon a tract of granite, such agents here work scarcely an appreciable effect; there cause exfoliations of the surface, and a resulting heap of *débris* and boulders; and elsewhere, after decomposing the feldspar into a white clay, carry away this with the accompanying quartz and mica, and deposit them in separate beds, fluviatile and marine. When the exposed land consists of several unlike formations, sedimentary and igneous, changes proportionably more heterogeneous are wrought. The formations being disintegrable in different degrees, there follows an increased irregularity of surface. The areas drained by different rivers being differently constituted, these rivers carry down to the sea unlike combinations of ingredients; and so sundry new strata of distinct composition arise. And here indeed we may see very simply illustrated, the truth, that the heterogeneity of the effects increases in a geometrical progression, with the heterogeneity of the object acted upon. A continent of complex structure, presenting many strata irregularly distributed, raised to various levels, tilted up at all angles, must, under the same denuding agencies, give origin to immensely multiplied results: each district must be peculiarly modified; each river must carry down a distinct kind of detritus; each deposit must be differently distributed by the entangled currents, tidal and other, which wash the contorted shores; and every additional complication of surface must be the cause of more than one additional consequence. But not to dwell on these, let us, for the fuller elucidation of this truth in relation to the inorganic world, consider what would presently follow from some extensive cosmical revolution—say the subsidence of Central America. The immediate results of the disturbance would themselves be sufficiently complex. Besides the numberless dislocations of strata, the ejections of igneous matter, the propagation of earthquake vibrations thousands of miles around, the loud explosions, and the escape of gases, there would be the rush of the Atlantic and Pacific Oceans to supply the vacant space, the subsequent recoil of enormous waves, which would traverse both these oceans and produce myriads of changes along their shores, the corresponding atmospheric waves complicated by the currents surrounding each volcanic vent, and the electrical discharges with which such disturbances are accompanied. But these temporary effects would be insignificant compared with the permanent ones. The complex currents of the Atlantic and Pacific would be altered in directions and amounts. The distribution of heat achieved by these currents would be different from what it is. The arrangement of the isothermal lines, not only on the neighbouring continents, but even throughout Europe, would be changed. The tides would flow differently from what they do now. There would be more or less modification of the winds in their periods, strengths, directions, qualities. Rain would fall scarcely anywhere at the same times and in the same quantities as at present. In short, the meteorological conditions thousands of miles off, on all sides, would be more or less revolutionized. In these many changes, each of which comprehends countless minor ones, the reader will see the immense heterogeneity of the results wrought out by one force, when that force expends itself on a previously complicated area; and he will readily draw the corollary that from the beginning the complication has advanced at an increasing rate.

§ 119. We have next to trace throughout organic evolution, this same all-pervading principle. And here, where the transformation of the homogeneous into the heterogeneous was first observed, the production of many changes by one cause is least easy to demonstrate. The development of a seed into a plant, or an ovum into an animal, is so gradual; while the forces which determine it are so involved, and at the same time so unobtrusive; that it is difficult to detect the multiplication of effects which is elsewhere so obvious. Nevertheless, by indirect evidence we may establish our proposition; spite of the lack of direct evidence.

Observe, first, how numerous are the changes which any marked stimulus works on an adult organism—a human being, for instance. An alarming sound or sight, besides impressions on the organs of sense and the nerves, may produce a start, a scream, a distortion of the face, a trembling consequent on general muscular relaxation, a burst of perspiration, an excited action of the heart, a rush of blood to the brain, followed possibly by arrest of the heart's action and by syncope; and if the system be feeble, an illness with its long train of complicated symptoms may set in. Similarly in cases of disease. A minute portion of the small-pox virus introduced into the system, will, in a severe case, cause, during the first stage, rigors, heat of skin, accelerated pulse, furred tongue, loss of appetite, thirst, epigastric uneasiness, vomiting, headache, pains in the back and limbs, muscular weakness, convulsions, delirium, &c.; in the second stage, cutaneous eruption, itching, tingling, sore throat, swelled fauces, salivation, cough, hoarseness, dyspnœa, &c.; and in the third stage, œdematous inflammations, pneumonia, pleurisy, diarrhœa, inflammation of the brain, ophthalmia, erysipelas, &c.: each of which enumerated symptoms is itself more or less complex. Medicines, special foods, better air, might in like manner be instanced as producing multiplied results. Now it needs only to consider that the many changes thus wrought by one force on an adult organism, must be partially paralleled in an embryo-organism, to understand how here also the production of many effects by one cause is a source of increasing heterogeneity. The external heat and other agencies which determine the first complications of the germ, will, by acting on these, superinduce further complications; on these still higher and more numerous ones; and so on continually: each organ as it is developed, serving, by its actions and reactions on the rest, to initiate new complexities. The first pulsations of the fœtal heart must simultaneously aid the unfolding of every part. The growth of each tissue, by taking from the blood special proportions of elements, must modify the constitution of the blood; and so must modify the nutrition of all the other tissues. The distributive actions, implying as they do a certain waste, necessitate an addition to the blood of effete matters, which must influence the rest of the system, and perhaps, as some think, initiate the formation of excretory organs. The nervous connections established among the viscera must further multiply their mutual influences. And so with every modification of structure—every additional part and every alteration in the ratios of parts. Still stronger becomes the proof when we call to mind the fact, that the same germ may be evolved into different forms according to circumstances. Thus, during its earlier stages, every embryo is sexless—becomes either male or female as the balance of forces acting on it determines. Again, it is well-known that the larva of a working-bee will develop into a queen-bee, if, before a certain period, its food be changed to that on which the larvæ of queen-bees are fed. Even more remarkable is the case of certain entozoa. The ovum of a tape-worm, getting into the intestine of one animal, unfolds into the form of its parent; but if carried into other parts of the system, or into the intestine of some unlike animal, it becomes one of the sac-like creatures, called by naturalists *Cysticerci*, or *Cœnuri*, or *Echinococci*—creatures so extremely different from the tape-worm in aspect and structure, that only after careful investigations have they been proved to have the same origin. All which instances imply that each advance in embryonic complication results from the action of incident forces on the complication previously existing. Indeed, the now accepted doctrine of epigenesis necessitates the conclusion that organic evolution proceeds after this manner. For since it is proved that no germ, animal or vegetal, contains the slightest rudiment, trace, or indication of the future organism—since the microscope has shown us that the first process set up in every fertilized germ is a process of repeated spontaneous fissions, ending in the production of a mass of cells, not one of which exhibits any special character; there seems no alternative but to conclude that the partial organization at any moment subsisting in a growing embryo, is transformed by the agencies acting on it into the succeeding phase of organization, and this into the next, until, through ever-increasing complexities, the ultimate form is reached. Thus, though the subtlety of the forces and the slowness of the metamorphosis, prevent us from *directly* tracing the genesis of many changes by one cause, throughout the successive stages which every embryo passes through; yet, *indirectly*, we have strong evidence that this is a source of increasing heterogeneity. We have marked how multitudinous are the effects which a single agency may generate in an adult organism; that a like multiplication of effects must happen in the unfolding organism, we have inferred from sundry illustrative cases; further, it has been pointed out that the ability which like germs have to originate unlike forms, implies that the successive transformations result from the new changes superinduced on previous changes; and we have seen that structureless as every germ originally is, the development of an organism out of it is otherwise incomprehensible. Doubtless we are still in the dark respecting those mysterious properties which make the germ, when subject to fit influences, undergo the special changes beginning this series of transformations. All here contended is, that given a germ possessing these mysterious properties, the evolution of an organism from it depends, in part, on that multiplication of effects which we have seen to be a cause of evolution in general, so far as we have yet traced it.

When, leaving the development of single plants and animals, we pass to that of the Earth's flora and fauna, the course of the argument again becomes clear and simple. Though, as before admitted, the fragmentary facts Palæontology has accumulated, do not clearly warrant us in saying that, in the lapse of geologic time, there have been evolved more heterogeneous organisms, and more heterogeneous assemblages of organisms; yet we shall now see that there *must* ever have been a tendency towards these results. We shall find that the production of many effects by one cause, which, as already shown, has been all along increasing the physical heterogeneity of the Earth, has further necessitated an increasing heterogeneity in its flora and fauna, individually and collectively. An illustration will make this clear. Suppose that by a series of upheavals, occurring, as they are now known to do, at long intervals, the East Indian Archipelago were to be raised into a continent, and a chain of mountains formed along the axis of elevation. By the first of these upheavals, the plants and animals inhabiting Borneo, Sumatra, New Guinea, and the rest, would be subjected to slightly-modified sets of conditions. The climate in general would be altered in temperature, in humidity, and in its periodical variations; while the local differences would be multiplied. These modifications would affect, perhaps inappreciably, the entire flora and fauna of the region. The change of level would produce additional modifications; varying in different species, and also in different members of the same species, according to their distance from the axis of elevation. Plants, growing only on the sea-shore in special localities, might become extinct. Others, living only in swamps of a certain humidity, would, if they survived at all, probably undergo visible changes of appearance. While more marked alterations would occur in some of the plants that spread over the lands newly raised above the sea. The animals and insects living on these modified plants, would themselves be in some degree modified by change of food, as well as by change of climate; and the modification would be more marked where, from the dwindling or disappearance of one kind of plant, an allied kind was eaten. In the lapse of the many generations arising before the next upheaval, the sensible or insensible alterations thus produced in each species, would become organized—in all the races that survived there would be a more or less complete adaptation to the new conditions. The next upheaval would superinduce further organic changes, implying wider divergences from the primary forms; and so repeatedly. Now however let it be observed that this revolution would not be a substitution of a thousand modified species for the thousand original species; but in place of the thousand original species there would arise several thousand species, or varieties, or changed forms. Each species being distributed over an area of some extent, and tending continually to colonize the new area exposed, its different members would be subject to different sets of changes. Plants and animals migrating towards the equator would not be affected in the same way with others migrating from it. Those which spread towards the new shores, would undergo changes unlike the changes undergone by those which spread into the mountains. Thus, each original race of organisms would become the root from which diverged several races, differing more or less from it and from each other; and while some of these might subsequently disappear, probably more than one would survive in the next geologic period: the very dispersion itself increasing the chances of survival. Not only would there be certain modifications thus caused by changes of physical conditions and food; but also in some cases other modifications caused by changes of habit. The fauna of each island, peopling, step by step, the newly-raised tracts, would eventually come in contact with the faunas of other islands; and some members of these other faunas would be unlike any creatures before seen. Herbivores meeting with new beasts of prey, would, in some cases, be led into modes of defence or escape differing from those previously used; and simultaneously the beasts of prey would modify their modes of pursuit and attack. We know that when circumstances demand it, such changes of habit *do* take place in animals; and we know that if the new habits become the dominant ones, they must eventually in some degree alter the organization.

Observe now, however, a further consequence. There must arise not simply a tendency towards the differentiation of each race of organisms into several races; but also a tendency to the occasional production of a somewhat higher organism. Taken in the mass, these divergent varieties, which have been caused by fresh physical conditions and habits of life, will exhibit alterations quite indefinite in kind and degree; and alterations that do not necessarily constitute an advance. Probably in most cases the modified type will be not appreciably more heterogeneous than the original one. But it *must* now and then occur, that some division of a species, falling into circumstances which give it rather more complex experiences, and demand actions somewhat more involved, will have certain of its organs further differentiated in proportionately small degrees—will become slightly more heterogeneous. Hence, there will from time to time arise an increased heterogeneity both of the Earth's flora and fauna, and of individual races included in them. Omitting detailed explanations, and allowing for the qualifications which cannot here be specified, it is sufficiently clear that geological mutations have all along tended to complicate the forms of life, whether regarded separately or collectively. That multiplication of effects which has been a part-cause of the transformation of the Earth's crust from the simple into the complex, has simultaneously led to a parallel transformation of the Life upon its surface.[17]

The deduction here drawn from the established truths of geology and the general laws of life, gains immensely in weight on finding it to be in harmony with an induction drawn from direct experience. Just that divergence of many races from one race, which we inferred must have been continually occurring during geologic time, we know to have occurred during the pre-historic and historic periods, in man and domestic animals. And just that multiplication of effects which we concluded must have been instrumental to the first, we see has in a great measure wrought the last. Single causes, as famine, pressure of population, war, have periodically led to further dispersions of mankind and of dependent creatures: each such dispersion initiating new modifications, new varieties of type. Whether all the human races be or be not derived from one stock, philology makes it clear that whole groups of races, now easily distinguishable from each other, were originally one race—that the diffusion of one race into different climates and conditions of existence has produced many altered forms of it. Similarly with domestic animals. Though in some cases (as that of dogs) community of origin will perhaps be disputed, yet in other cases (as that of the sheep or the cattle of our own country) it will not be questioned that local differences of climate, food, and treatment, have transformed one original breed into numerous breeds, now become so far distinct as to produce unstable hybrids. Moreover, through the complication of effects flowing from single causes, we here find, what we before inferred, not only an increase of general heterogeneity, but also of special heterogeneity. While of the divergent divisions and subdivisions of the human race, many have undergone changes not constituting an advance; others have become decidedly more heterogeneous. The civilized European departs more widely from the vertebrate archetype than does the savage.

§ 120. A sensation does not expend itself in arousing some single state of consciousness; but the state of consciousness aroused is made up of various represented sensations connected by co-existence, or sequence with the presented sensation. And that, in proportion as the grade of intelligence is high, the number of ideas suggested is great, may be readily inferred. Let us, however, look at the proof that here too, each change is the parent of many changes; and that the multiplication increases in proportion as the area affected is complex.

Were some hitherto unknown bird, driven say by stress of weather from the remote north, to make its appearance on our shores, it would excite no speculation in the sheep or cattle amid which it alighted: a perception of it as a creature like those constantly flying about, would be the sole interruption of that dull current of consciousness which accompanies grazing and rumination. The cow-herd, by whom we may suppose the exhausted bird to be presently caught, would probably gaze at it with some slight curiosity, as being unlike any he had before seen—would note its most conspicuous markings, and vaguely ponder on the questions, where it came from, and how it came. The village bird-stuffer would have suggested to him by the sight of it, sundry forms to which it bore a little resemblance; would receive from it more numerous and more specific impressions respecting structure and plumage; would be reminded of various instances of birds brought by storms from foreign parts—would tell who found them, who stuffed them, who bought them. Supposing the unknown bird taken to a naturalist of the old school, interested only in externals, (one of those described by the late Edward Forbes, as examining animals as though they were merely skins filled with straw,) it would excite in him a more involved series of mental changes: there would be an elaborate examination of the feathers, a noting of all their technical distinctions, with a reduction of these perceptions to certain equivalent written symbols; reasons for referring the new form to a particular family, order, and genus would be sought out and written down; communications with the secretary of some society, or editor of some journal, would follow; and probably there would be not a few thoughts about the addition of the *ii* to the describer's name, to form the name of the species. Lastly, in the mind of a comparative anatomist, such a new species, should it happen to have any marked internal peculiarity, might produce additional sets of changes—might very possibly suggest modified views respecting the relationships of the division to which it belonged; or, perhaps, alter his conceptions of the homologies and developments of certain organs; and the conclusions drawn might not improbably enter as elements into still wider inquiries concerning the origin of organic forms.

From ideas let us turn to emotions. In a young child, a father's anger produces little else than vague fear—a disagreeable sense of impending evil, taking various shapes of physical suffering or deprivation of pleasures. In elder children, the same harsh words will arouse additional feelings: sometimes a sense of shame, of penitence, or of sorrow for having offended; at other times, a sense of injustice, and a consequent anger. In the wife, yet a further range of feelings may come into existence—perhaps wounded affection, perhaps self-pity for ill-usage, perhaps contempt for groundless irritability, perhaps sympathy for some suffering which the irritability indicates, perhaps anxiety about an unknown misfortune which she thinks has produced it. Nor are we without evidence that among adults, the like differences of development are accompanied by like differences in the number of emotions that are aroused, in combination or rapid succession— the lower natures being characterized by that impulsiveness which results from the uncontrolled action of a few feelings; and the higher natures being characterized by the simultaneous action of many secondary feelings, modifying those first awakened.

Possibly it will be objected that the illustrations here given, are drawn from the functional changes of the nervous system, not from its structural changes; and that what is proved among the first, does not necessarily hold among the last. This must be admitted. Those, however, who recognize the truth that the structural changes are the slowly accumulated results of the functional changes, will readily draw the corollary, that a part-cause of the evolution of the nervous system, as of other evolution, is this multiplication of effects which becomes ever greater as the development becomes higher.

§ 121. If the advance of Man towards greater heterogeneity in both body and mind, is in part traceable to the production of many effects by one cause, still more clearly may the advance of Society towards greater heterogeneity be so explained. Consider the growth of an industrial organization. When, as must occasionally happen, some individual of a tribe displays unusual aptitude for making an article of general use (a weapon, for instance) which was before made by each man for himself, there arises a tendency towards the differentiation of that individual into a maker of weapons. His companions (warriors and hunters all of them) severally wish to have the best weapons that can be made; and are therefore certain to offer strong inducements to this skilled individual to make weapons for them. He, on the other hand, having both an unusual faculty, and an unusual liking, for making weapons (the capacity and the desire for any occupation being commonly associated), is predisposed to fulfil these commissions on the offer of adequate rewards: especially as his love of distinction is also gratified. This first specialization of function, once commenced, tends ever to become more decided. On the side of the weapon-maker, continued practice gives increased skill—increased superiority to his products. On the side of his clients, cessation of practice entails decreased skill. Thus the influences that determine this division of labour grow stronger in both ways: this social movement tends ever to become more decided in the direction in which it was first set up; and the incipient heterogeneity is, on the average of cases, likely to become permanent for that generation, if no longer. Such a process, besides differentiating the social mass into two parts, the one monopolizing, or almost monopolizing, the performance of a certain function, and the other having lost the habit, and in some measure the power, of performing that function, has a tendency to initiate other differentiations. The advance described implies the introduction of barter: the maker of weapons has, on each occasion, to be paid in such other articles as he agrees to take in exchange. Now he will not habitually take in exchange one kind of article, but many kinds. He does not want mats only, or skins, or fishing-gear; but he wants all these; and on each occasion will bargain for the particular things he most needs. What follows? If among the members of the tribe there exist any slight differences of skill in the manufacture of these various things, as there are almost sure to do, the weapon-maker will take from each one the thing which that one excels in making: he will exchange for mats with him whose mats are superior, and will bargain for the fishing-gear of whoever has the best. But he who has bartered away his mats or his fishing-gear, must make other mats or fishing-gear for himself; and in so doing must, in some degree, further develop his aptitude. Thus it results that the small specialities of faculty possessed by various members of the tribe will tend to grow more decided. If such transactions are from time to time repeated, these specializations may become appreciable. And whether or not there ensue distinct differentiations of other individuals into makers of particular articles, it is clear that incipient differentiations take place throughout the tribe: the one original cause produces not only the first dual effect, but a number of secondary dual effects, like in kind but minor in degree. This process, of which traces may be seen among groups of school-boys, cannot well produce a lasting distribution of functions in an unsettled tribe; but where there grows up a fixed and multiplying community, such differentiations become permanent, and increase with each generation. An addition to the number of citizens, involving a greater demand for every commodity, intensifies the functional activity of each specialized person or class; and this renders the specialization more definite where it already exists, and establishes it where it is but nascent. By increasing the pressure on the means of subsistence, a larger population again augments these results; since every individual is forced more and more to confine himself to that which he can do best, and by which he can gain most. And this industrial progress, by aiding future production, opens the way for further growth of population, which reacts as before. Presently, under the same stimuli, new occupations arise. Competing workers, severally aiming to produce improved articles, occasionally discover better processes or better materials. In weapons and cutting-tools, the substitution of bronze for stone entails on him who first makes it, a great increase of demand—so great an increase that he presently finds all his time occupied in making the bronze for the articles he sells, and is obliged to depute the fashioning of these articles to others; and eventually the making of bronze, thus gradually differentiated from a pre-existing occupation, becomes an occupation by itself. But now mark the ramified changes which follow this change. Bronze soon replaces stone, not only in the articles it was first used for, but in many others; and so affects the manufacture of them. Further, it affects the processes which such improved utensils subserve, and the resulting products—modifies buildings, carvings, dress, personal decorations. Yet again, it sets going sundry manufactures which were before impossible, from lack of a material fit for the requisite tools. And all these changes react on the people—increase their manipulative skill, their intelligence, their comfort—refine their habits and tastes.

It is out of the question here to follow through its successive complications, this increasing social heterogeneity that results from the production of many effects by one cause. But leaving the intermediate phases of social development, let us take an illustration from its passing phase. To trace the effects of steam-power, in its manifold applications to mining, navigation, and manufactures, would carry us into unmanageable detail. Let us confine ourselves to the latest embodiment of steam-power— the locomotive engine. This, as the proximate cause of our railway-system, has changed the face of the country, the course of trade, and the habits of the people. Consider, first, the complicated sets of changes that precede the making of every railway—the provisional arrangements, the meetings, the registration, the trial-section, the parliamentary survey, the lithographed plans, the books of reference, the local deposits and notices, the application to Parliament, the passing Standing-Orders Committee, the first, second, and third readings: each of which brief heads indicates a multiplicity of transactions, and the further development of sundry occupations, (as those of engineers, surveyors, lithographers, parliamentary agents, share-brokers,) and the creation of sundry others (as those of traffic-takers, reference-takers). Consider, next, the yet more marked changes implied in railway construction—the cuttings, em-bankings, tunnellings, diversions of roads; the building of bridges and stations; the laying down of ballast, sleepers, and rails; the making of engines, tenders, carriages, and wagons: which processes, acting upon numerous trades, increase the importation of timber, the quarrying of stone, the manufacture of iron, the mining of coal, the burning of bricks; institute a variety of special manufactures weekly advertised in the *Railway Times*; and call into being some new classes of workers—drivers, stokers, cleaners, plate-layers, &c. &c. Then come the changes, more numerous and involved still, which railways in action produce on the community at large. The organization of every business is more or less modified: ease of communication makes it better to do directly what was before done by proxy; agencies are established where previously they would not have paid; goods are obtained from remote wholesale houses instead of near retail ones; and commodities are used which distance once rendered inaccessible. The rapidity and small cost of carriage, tend to specialize more than ever the industries of different districts—to confine each manufacture to the parts in which, from local advantages, it can be best carried on. Economical distribution equalizes prices, and also, on the average, lowers prices: thus bringing divers articles within the means of those before unable to buy them, and so increasing their comforts and improving their habits. At the same time the practice of travelling is immensely extended. Classes who before could not afford it, take annual trips to the sea; visit their distant relations; make tours; and so we are benefited in body, feelings, and intellect. The more prompt transmission of letters and of news produces further changes—makes the pulse of the nation faster. Yet more, there arises a wide dissemination of cheap literature through railway book-stalls, and of advertisements in railway carriages: both of them aiding ulterior progress. And the innumerable changes here briefly indicated are consequent on the invention of the locomotive engine. The social organism has been rendered more heterogeneous, in virtue of the many new occupations introduced, and the many old ones further specialized; prices in all places have been altered; each trader has, more or less, modified his way of doing business; and every person has been affected in his actions, thoughts, emotions.

The only further fact demanding notice, is, that we here see more clearly than ever, that in proportion as the area over which any influence extends, becomes heterogeneous, the results are in a yet higher degree multiplied in number and kind. While among the primitive tribes to whom it was first known, caoutchouc caused but few changes, among ourselves the changes have been so many and varied that the history of them occupies a volume. Upon the small, homogeneous community inhabiting one of the Hebrides, the electric telegraph would produce, were it used, scarcely any results; but in England the results it produces are multitudinous.

Space permitting, the synthesis might here be pursued in relation to all the subtler products of social life. It might be shown how, in Science, an advance of one division presently advances other divisions—how Astronomy has been immensely forwarded by discoveries in Optics, while other optical discoveries have initiated Microscopic Anatomy, and greatly aided the growth of Physiology—how Chemistry has indirectly increased our knowledge of Electricity, Magnetism, Biology, Geology—how Electricity has reacted on Chemistry and Magnetism, developed our views of Light and Heat, and disclosed sundry laws of nervous action. In Literature the same truth might be exhibited in the still-multiplying forms of periodical publications that have descended from the first newspaper, and which have severally acted and reacted on other forms of literature and on each other; or in the bias given by each book of power to various subsequent books. The influence which a new school of Painting (as that of the pre-Raphaelites) exercises on other schools; the hints which all kinds of pictorial art are deriving from Photography; the complex results of new critical doctrines; might severally be dwelt on as displaying the like multiplication of effects. But it would needlessly tax the reader's patience to detail, in their many ramifications, these various changes: here become so involved and subtle as to be followed with some difficulty.

§ 122. After the argument which closed the last chapter, a parallel one seems here scarcely required. For symmetry's sake, however, it will be proper briefly to point out how the multiplication of effects, like the instability of the homogeneous, is a corollary from the persistence of force.

Things which we call different are things which react in different ways; and we can know them as different only by the differences in their reactions. When we distinguish bodies as hard and soft, rough and smooth, we simply mean that certain like muscular forces expended on them are followed by unlike sets of sensations—unlike re-active forces. Objects that are classed as red, blue, yellow, &c., are objects that decompose light in strongly-contrasted ways; that is, we know contrasts of colour as contrasts in the changes produced in a uniform incident force. Manifestly, any two things which do not work unequal effects on consciousness, either by unequally opposing our own energies, or by impressing our senses with unequally modified forms of certain external energies, cannot be distinguished by us. Hence the proposition that the different parts of any whole must react differently on a uniform incident force, and must so reduce it to a group of multiform forces, is in essence a truism. A further step will reduce this truism to its lowest terms.

When, from unlikeness between the effects they produce on consciousness, we predicate unlikeness between two objects, what is our warrant? and what do we mean by the unlikeness, objectively considered? Our warrant is the persistence of force. Some kind or amount of change has been wrought in us by the one, which has not been wrought by the other. This change we ascribe to some force exercised by the one which the other has not exercised. And we have no alternative but to do this, or to assert that the change had no antecedent; which is to deny the persistence of force. Whence it is further manifest that what we regard as the objective unlikeness is the presence in the one of some force, or set of forces, not present in the other—something in the kinds or amounts or directions of the constituent forces of the one, which those of the other do not parallel. But now if things or parts of things which we call different, are those of which the constituent forces differ in one or more respects; what must happen to any like forces, or any uniform force, falling on them? Such like forces, or parts of a uniform force, must be differently modified. The force which is present in the one and not in the other, must be an element in the conflict— must produce its equivalent reaction; and must so affect the total reaction. To say otherwise is to say that this differential force will produce no effect; which is to say that force is not persistent.

I need not develop this corollary further. It manifestly follows that a uniform force, falling on a uniform aggregate, must undergo dispersion; that falling on an aggregate made up of unlike parts, it must undergo dispersion from each part, as well as qualitative differentiations; that in proportion as the parts are unlike, these qualitative differentiations must be marked; that in proportion to the number of the parts, they must be numerous; that the secondary forces so produced, must undergo further transformations while working equivalent transformations in the parts that change them; and similarly with the forces they generate. Thus the conclusions that a part-cause of Evolution is the multiplication of effects; and that this increases in geometrical progression as the heterogeneity becomes greater; are not only to be established inductively, but are deducible from the deepest of all truths.

> 17. Had this paragraph, first published in the *Westminster Review* in 1857, been written after the appearance of Mr. Darwin's work on *The Origin of Species*, it would doubtless have been otherwise expressed. Reference would have

been made to the process of "natural selection," as greatly facilitating the differentiations described. As it is, however, I prefer to let the passage stand in its original shape: partly because it seems to me that these successive changes of conditions would produce divergent varieties or species, apart from the influence of "natural selection" (though in less numerous ways as well as less rapidly); and partly because I conceive that in the absence of these successive changes of conditions, "natural selection" would effect comparatively little. Let me add that though these positions are not enunciated in *The Origin of Species*, yet a mutual friend gives me reason to think that Mr. Darwin would coincide in them; if he did not indeed consider them as tacitly implied in his work.

CHAPTER XV
DIFFERENTIATION AND INTEGRATION

§ 123. The general interpretation of Evolution is far from being completed in the preceding chapters. We must contemplate its changes under yet another aspect, before we can form a definite conception of the process constituted by them. Though the laws already set forth, furnish a key to the re-arrangement of parts which Evolution exhibits, in so far as it is an advance from the uniform to the multiform; they furnish no key to this re-arrangement in so far as it is an advance from the indefinite to the definite. On studying the actions and re-actions everywhere going on, we have found it to follow inevitably from a certain primordial truth, that the homogeneous must lapse into the heterogeneous, and that the heterogeneous must become more heterogeneous; but we have not discovered why the differently-affected parts of any simple whole, become clearly marked off from each other, at the same time that they become unlike. Thus far no reason has been assigned why there should not ordinarily arise a vague chaotic heterogeneity, in place of that orderly heterogeneity displayed in Evolution. It still remains to find out the cause of that integration of parts which accompanies their differentiation—that gradually-completed segregation of like units into a group, distinctly separated from neighbouring groups which are severally made up of other kinds of units. The rationale will be conveniently introduced by a few instances in which we may watch this segregative process taking place.

When towards the end of September, the trees are gaining their autumn colours, and we are hoping shortly to see a further change increasing still more the beauty of the landscape, we are not uncommonly disappointed by the occurrence of an equinoxial gale. Out of the mixed mass of foliage on each branch, the strong current of air carries away the decaying and brightly-tinted leaves, but fails to detach those which are still green. And while these last, frayed and seared by long-continued beatings against each other, and the twigs around them, give a sombre colour to the woods, the red and yellow and orange leaves are collected together in ditches and behind walls and in corners where eddies allow them to settle. That is to say, by the action of that uniform force which the wind exerts on both kinds, the dying leaves are picked out from among their still living companions and gathered in places by themselves. Again, the separation of particles of different sizes, as dust and sand from pebbles, may be similarly effected; as we see on every road in March. And from the days of Homer downwards, the power of currents of air, natural and artificial, to part from one another units of unlike specific gravities, has been habitually utilized in the winnowing of chaff from wheat. In every river we see how the mixed materials carried down, are separately deposited—how in rapids the bottom gives rest to nothing but boulders and pebbles; how where the current is not so strong, sand is let fall; and how, in still places, there is a sediment of mud. This selective action of moving water, is commonly applied in the arts to obtain masses of particles of different degrees of fineness. Emery, for example, after being ground, is carried by a slow current through successive compartments; in the first of which the largest grains subside; in the second of which the grains that reach the bottom before the water has escaped, are somewhat smaller; in the third smaller still; until in the last there are deposited only those finest particles which fall so slowly through the water, that they have not previously been able to reach the bottom. And in a way that is different though equally significant, this segregative effect of water in motion, is exemplified in the carrying away of soluble from insoluble matters—an application of it hourly made in every laboratory. The effects of the uniform forces which aerial and aqueous currents exercise, are paralleled by those of uniform forces of other orders. Electric attraction will separate small bodies from large, or light bodies from heavy. By magnetism, grains of iron may be selected from among other grains; as by the Sheffield grinder, whose magnetized gauze mask filters out the steel-dust which his wheel gives off, from the stone-dust that accompanies it. And how the affinity of any agent acting differently on the components of a given body, enables us to take away some component and leave the rest behind, is shown in almost every chemical experiment.

What now is the general truth here variously presented? How are these several facts and countless similar ones, to be expressed in terms that embrace them all? In each case we see in action a force which may be regarded as simple or uniform—fluid motion in a certain direction at a certain velocity; electric or magnetic attraction of a given amount; chemical affinity of a particular kind: or rather, in strictness, the acting force is compounded of one of these and certain other uniform forces, as gravitation, etc. In each case we have an aggregate made up of unlike units—either atoms of different substances combined or intimately mingled, or fragments of the same substance of different sizes, or other constituent parts that are unlike in their specific gravities, shapes, or other attributes. And in each case these unlike units, or groups of units, of which the aggregate consists, are, under the influence of some resultant force acting indiscriminately on them all, separated from each other—segregated into minor aggregates, each consisting of units that are severally like each other and unlike those of the other minor aggregates. Such being the common aspect of these changes, let us look for the common interpretation of them.

In the chapter on "The Instability of the Homogeneous," it was shown that a uniform force falling on any aggregate, produces unlike modifications in its different parts—turns the uniform into the multiform and the multiform into the more multiform. The transformation thus wrought, consists of either insensible or sensible changes of relative position among the units, or of both—either of those molecular re-arrangements which we call chemical, or of those larger transpositions which are distinguished as mechanical, or of the two united. Such portion of the permanently effective force as reaches each different part, or differently-conditioned part, may be expended in modifying the mutual relations of its constituents; or it may be expended in moving the part to another place; or it may be expended partially in the first and partially in the second. Hence, so much of the permanently effective force as does not work the one kind of effect, must work the other kind. It is manifest that if of the permanently effective force which falls on some compound unit of an aggregate, little, if any, is absorbed in re-arranging the ultimate components of such compound unit, much or the whole, must show itself in motion of such compound unit to some other place in the aggregate; and conversely, if little or none of this force is absorbed in generating mechanical transposition, much or the whole must go to produce molecular alterations. What now must follow from this? In cases where none or only part of the force generates chemical re-distributions, what physical re-distributions must be generated? Parts that are similar to each other will be similarly acted on by the force; and will similarly react on it. Parts that are dissimilar will be dissimilarly acted on by the force; and will dissimilarly react on it. Hence the permanently effective incident force, when wholly or partially transformed into mechanical motion of the units, will produce like motions in units that are alike, and unlike motions in units that are unlike. If then, in an aggregate containing two or more orders of mixed units, those of the same order will be moved in the same way, and in a way that differs from that in which units of other orders are moved, the respective orders must segregate. A group of like things on which are impressed motions that are alike in amount and direction, must be transferred as a group to another place, and if they are mingled with some group of other things, on which the motions impressed are like each other, but unlike those of the first group in amount or direction or both, these other things must be transferred as a group to some other place—the mixed aggregate must undergo a simultaneous differentiation and integration.

In further elucidation of this process, it will be well here to set down a few instances in which we may see that, other things equal, the definiteness of the separation is in proportion to the definiteness of the difference between the units. Take a handful of any pounded substance, containing fragments of all sizes; and let it fall to the ground while a gentle breeze is blowing. The large fragments will be collected together on the ground almost immediately under the hand; somewhat smaller fragments will be carried a little to the leeward; still smaller ones a little further; and those minute particles which we call dust, will be drifted a long way before they reach the earth: that is, the integration is indefinite where the difference among the fragments is indefinite, though the divergence is greatest where the difference is greatest. If, again, the handful be made up of quite distinct orders of units—as pebbles, coarse sand, and dust—these will, under like conditions, be segregated with comparative definiteness: the pebbles will drop almost vertically; the sand will fall in an inclined direction, and deposit itself within a tolerably circumscribed space beyond the pebbles; while the dust will be blown almost horizontally to a great distance. A case in which another kind of force comes into play, will still better illustrate this truth. Through a mixed aggregate of soluble and insoluble substances, let water slowly percolate. There will in the first place be a distinct parting of the substances that are the most widely contrasted in their relations to the acting forces: the soluble will be carried away; the insoluble will remain behind. Further, some separation, though a less definite one, will be effected among the soluble substances; since the first part of the current will remove the most soluble substances in the largest amounts, and after these have been all dissolved, the current will still continue to bring out the remaining less soluble substances. Even the undissolved matters will have simultaneously undergone a certain segregation; for the percolating fluid will carry down the minute fragments from among the large ones, and will deposit those of small specific gravity in one place, and those of great specific gravity in another. To complete the elucidation we must glance at the obverse fact; namely, that mixed units which differ but slightly, are moved in but slightly-different ways by incident forces, and can therefore be separated only by such adjustments of the incident forces as allow slight differences to become appreciable factors in the result. This truth is made manifest by antithesis in the instances just given; but it may be made much more manifest by a few such instances as those which chemical analysis supplies in abundance. The parting of alcohol from water by distillation is a good one. Here we have atoms consisting of oxygen and hydrogen, mingled with atoms consisting of oxygen, hydrogen, and carbon. The two orders of atoms have a considerable similarity of nature: they similarly maintain a fluid form at ordinary temperatures; they similarly become gaseous more and more rapidly as the temperature is raised; and they boil at points not very far apart. Now this comparative likeness of the atoms is accompanied by difficulty in segregating them. If the mixed fluid is unduly heated, much water distils over with the alcohol: it is only within a narrow range of temperature, that the one set of atoms are driven off rather than the others; and even then not a few of the others accompany them. The most interesting and instructive example, however, is furnished by certain phenomena of crystallization. When several salts that have little analogy of constitution, are dissolved in the same body of water, they are separated without much trouble, by crystallization: their respective units moved towards each other, as physicists suppose, by polar forces, segregate into crystals of their respective kinds. The crystals of each salt do, indeed, usually contain certain small amounts of the other salts present in the solution—especially when the crystallization has been rapid; but from these other salts they are severally freed by repeated resolutions and crystallizations. Mark now, however, that the reverse is the case when the salts contained in the same body of water are chemically homologous. The nitrates of baryta and lead, or the sulphates of zinc, soda, and magnesia, unite in the same crystals; nor will they crystallize separately if these crystals be dissolved afresh, and afresh crystallized, even with great care. On seeking the cause of this anomaly, chemists found that such salts were isomorphous—that their atoms, though not chemically identical, were identical in the proportions of acid, base, and water, composing them, and in their crystalline forms: whence it was inferred that their atoms are nearly alike in structure. Thus is clearly illustrated the truth, that units of unlike kinds are differentiated and integrated with a readiness proportionate to the degree of their unlikeness. In the first case we see that being dissimilar in their forms, but similar in so far as they are soluble in water of a certain temperature, the atoms segregate, though imperfectly. In the second case we see that the atoms, having not only the likeness implied by solubility in the same menstruum, but also a great likeness of structure, do not segregate—are differentiated and integrated only under quite special conditions, and then very incompletely. That is, the incident force of mutual polarity impresses unlike motions on the mixed units in proportion as they are unlike; and therefore, in proportion as they are unlike, tends to deposit them in separate places.

There is a converse cause of segregation, which it is needless here to treat of with equal fulness. If different units acted on by the same force, must be differently moved; so, too, must units of the same kind be differently moved by different forces. Supposing some group of units forming part of a homogeneous aggregate, are unitedly exposed to a force that is unlike in amount or direction to the force acting on the rest of the aggregate; then this group of units will separate from the rest, provided that, of the force so acting on it, there remains any portion not dissipated in molecular vibrations, nor absorbed in producing molecular re-arrangements. After all that has been said above, this proposition needs no defence.

Before ending our preliminary exposition, a complementary truth must be specified; namely, that mixed forces are segregated by the reaction of uniform matters, just as mixed matters are segregated by the action of uniform forces. Of this truth a complete and sufficient illustration is furnished by the dispersion of refracted light. A beam of light, made up of ethereal undulations of different orders, is not uniformly deflected by a homogeneous refracting body; but the different orders of undulations it contains, are deflected at different angles: the result being that these different orders of undulations are separated and integrated, and so produce what we know as the colours of the spectrum. A segregation of another kind occurs when rays of light traverse an obstructing medium. Those rays which consist of comparatively short undulations, are absorbed before those which consist of comparatively long ones; and the red rays, which consist of the longest undulations, alone penetrate when the obstruction is very great. How, conversely, there is produced a separation of like forces by the reaction of unlike matters, is also made manifest by the phenomena of refraction: since adjacent and parallel beams of light, falling on, and passing through, unlike substances, are made to diverge.

§ 124. On the assumption of their nebular origin, stars and planets exemplify that cause of material integration last assigned—the action of unlike forces on like units.

In a preceding chapter (§ 110) we saw that if matter ever existed in a diffused form, it could not continue uniformly distributed, but must break up into masses. It was shown that in the absence of a perfect balance of mutual attractions among atoms dispersed through unlimited space, there must arise breeches of continuity throughout the aggregate formed by them, and a concentration of it towards centres of dominant attraction. Where any such breech of continuity occurs, and the atoms that were before adjacent separate from each other; they do so in consequence of a difference in the forces to which they are respectively subject. The atoms on the one side of the breech are exposed to a certain surplus attraction in the direction in which they begin to move; and those on the other to a surplus attraction in the opposite direction. That is, the adjacent groups of like units are exposed to unlike resultant forces; and accordingly separate and integrate.

The formation and detachment of a nebulous ring, illustrates the same general principle. To conclude, as Laplace did, that the equatorial portion of a rotating nebulous spheroid, will, during concentration, acquire a centrifugal force sufficient to prevent it from following the rest of the contracting mass, is to conclude that such portions will remain behind as are in common subject to a certain differential force. The line of division between the ring and the spheroid, must be a line inside of which the aggregative force is greater than the force resisting aggregation; and outside of which the force resisting aggregation is greater than the aggregative force. Hence the alleged process conforms to the law that among like units, separation and integration is produced by the action of unlike forces.

Astronomical phenomena do not furnish any other than these hypothetical examples. In its present comparatively settled condition, the Solar System exhibits no direct evidence of progressing integration: unless indeed under the insignificant form of the union of meteoric masses with the Earth, and, occasionally perhaps, of cometary matter with the Sun.

§ 125. Those geologic changes usually classed as aqueous, display under numerous forms the segregation of unlike units by a uniform incident force. On sea-shores, the waves are ever sorting-out and separating the mixed materials against which they break. From each mass of fallen cliff, the rising and ebbing tide carries away all those particles which are so small as to remain long suspended in the water; and, at some distance from shore, deposits them in the shape of fine sediment. Large particles, sinking with comparative rapidity, are accumulated into beds of sand near low water-mark. The coarse grit and small pebbles collect together on the incline up which the breakers rush. And on the top lie the larger stones and boulders. Still more specific segregations may occasionally be observed. Flat pebbles, produced by the breaking down of laminated rock, are sometimes separately collected in one part of a shingle bank. On this shore the deposit is wholly of mud; on that it is wholly of sand. Here we find a sheltered cove filled with small pebbles almost of one size; and there, in a curved bay one end of which is more exposed than the other, we see a progressive increase in the massiveness of the stones as we walk from the less exposed to the more exposed end. Indeed, our sedimentary strata form one vast series of illustrations of the alleged law. Trace the history of each deposit, and we are quickly led down to the fact, that mixed fragments of matter, differing in their sizes or weights, are, when exposed to the momentum and friction of water, joined with the attraction of the Earth, selected from each other, and united into groups of comparatively like fragments. We see that, other things equal, the separation is definite in proportion as the differences of the units are marked; and that, under the action of the same aggregate of forces, the most widely unlike units are most widely removed from each other.

Among igneous changes we do not find so many examples of the process described. When specifying the conditions to Evolution, it was pointed out (§ 104) that molecular vibration exceeding a certain intensity, does not permit those integrations which result from the action of minor differential forces. Nevertheless, geological phenomena of this order are not barren of illustrations. Where the mixed matters composing the Earth's crust have been raised to a very high temperature, segregation habitually takes place as the temperature diminishes. Sundry of the substances that escape in a gaseous form from volcanoes, sublime into crystals on coming against cool surfaces; and solidifying, as these substances do, at different temperatures, they are deposited at different parts of the crevices through which they are emitted together. The best illustration, however, is furnished by the changes that occur during the slow cooling of igneous rock. When, through one of the fractures from time to time made in the solid shell which forms the Earth's crust, a portion of the molten nucleus is extruded; and when this is cooled with comparative rapidity, through free radiation and contact with cold masses; it forms a substance known as trap or basalt—a substance that is uniform in texture, though made up of various ingredients. But when, not escaping through the superficial strata, such a portion of the molten nucleus is slowly cooled, it becomes what we know as granite: the mingled particles of quartz, feldspar, and mica, being kept for a long time in a fluid and semi-fluid state—a state of comparative mobility—undergo those changes of position which the forces impressed on them by their fellow units necessitate. Having time in which to generate the requisite motions of the atoms, the differential forces arising from mutual polarity, segregate the quartz, feldspar, and mica, into crystals. How completely this is dependent on the long-continued agitation of the mixed particles, and consequent long-continued mobility by small differential forces, is proved by the fact that in granite dykes, the crystals in the centre of the mass, where the fluidity or semi-fluidity continued for a longer time, are much larger than those at the sides, where contact with the neighbouring rock caused more rapid cooling and solidification.

§ 126. The actions going on throughout an organism are so involved and subtle, that we cannot expect to identify the particular forces by which particular integrations are effected. Among the few instances admitting of tolerably definite interpretation, the best are those in which mechanical pressures and tensions are the agencies at work. We shall discover several on studying the bony frame of the higher animals.

The vertebral column of a man, is subject, as a whole, to certain general strains—the weight of the body, together with the reactions involved by all considerable muscular efforts; and in conformity with this, it has a certain general integration. At the same time, being exposed to different forces in the course of those lateral bendings which the movements necessitate, its parts retain a certain separateness. And if we trace up the development of the vertebral column from its primitive form of a cartilaginous cord in the lowest fishes, we see that, throughout, it maintains an integration corresponding to the unity of the incident forces, joined with a division into segments corresponding to the variety of the incident forces. Each segment, considered apart, exemplifies the truth more simply. A vertebra is not a single bone, but consists of a central mass with sundry appendages or processes; and in rudimentary types of vertebræ, those appendages are quite separate from the central mass, and, indeed, exist before it makes its appearance. But these several independent bones, constituting a primitive spinal segment, are subject to a certain aggregate of forces which agree more than they differ: as the fulcrum to a group of muscles habitually acting together, they perpetually undergo certain reactions in common. And accordingly, we see that in the course of development they gradually coalesce. Still clearer is the illustration furnished by spinal segments that become fused together where they are together exposed to some predominant strain. The sacrum consists of a group of vertebræ firmly united. In the ostrich and its congeners there are from seventeen to twenty sacral vertebræ; and besides being confluent with each other, these are confluent with the iliac bones, which run on each side of them. If now we assume these vertebræ to have been originally separate, as they still are in the embryo bird; and if we consider the mechanical conditions to which they must in such case have been exposed; we shall see that their union results in the alleged way. For through these vertebræ the entire weight of the body is transferred to the legs: the legs support the pelvic arch; the pelvic arch supports the sacrum; and to the sacrum is articulated the rest of the spine, with all the limbs and organs attached to it. Hence, if separate, the sacral vertebræ must be held firmly together by strongly-contracted muscles; and must, by implication, be prevented from partaking in those lateral movements which the other vertebræ undergo—they must be subject to a common strain, while they are preserved from strains which would affect them differently; and so they fulfil the conditions under which integration occurs. But the cases in which cause and effect are brought into the most obvious relation, are supplied by the limbs. The metacarpal bones (those which in man support the palm of the hand) are separate from each other in the majority of mammalia: the separate actions of the toes entailing on them slight amounts of separate movements. This is not so however in the ox-tribe and the horse-tribe. In the ox-tribe, only the middle metacarpals (third and fourth) are developed; and these, attaining massive proportions, coalesce to form the cannon bone. In the horse-tribe, the integration is what we may distinguish as indirect: the second and fourth metacarpals are present only as rudiments united to the sides of the third, while the third is immensely developed; thus forming a cannon bone which differs from that of the ox in being a single cylinder, instead of two cylinders fused together. The metatarsus in these quadrupeds exhibits parallel changes. Now each of these metamorphoses occurs where the different bones grouped together have no longer any different functions, but retain only a common function. The feet of oxen and horses are used solely for locomotion—are not put like those of unguiculate mammals to purposes which involve some relative movements of the metacarpals. Thus there directly or indirectly results a single mass of bone where the incident force is single. And for the inference that these facts have a causal connexion, we find confirmation throughout the entire class of birds; in the wings and legs of which, like integrations are found under like conditions. While this sheet is passing through the press, a fact illustrating this general truth in a yet more remarkable manner, has been mentioned to me by Prof. Huxley; who kindly allows me to make use of it while still unpublished by him. The *Glyptodon*, an extinct mammal found fossilized in South America, has long been known as a large uncouth creature allied to the Armadillo, but having a massive dermal armour consisting of polygonal plates closely fitted together so as to make a vast box, inclosing the body in such way as effectually to prevent it from being bent, laterally or vertically, in the slightest degree. This bony box, which must have weighed several hundred-weight, was supported on the spinous processes of the vertebræ, and on the adjacent bones of the pelvic and thoracic arches. And the significant fact now to be noted, is, that here, where the trunk vertebræ were together exposed to the pressure of this heavy dermal armour, at the same time that, by its rigidity, they were preserved from all relative movements, the entire series of them were united into one solid, continuous bone.

The formation and maintenance of a species, considered as an assemblage of similar organisms, is interpretable in an analogous way. We have already seen that in so far as the members of a species are subject to different sets of incident forces, they are differentiated, or divided into varieties. And here it remains to add that in so far as they are subject to like sets of incident forces, they are integrated, or reduced to, and kept in, the state of a uniform aggregate. For by the process of "natural selection," there is a continual purification of each species from those individuals which depart from the common type in ways that unfit them for the conditions of their existence. Consequently, there is a continual leaving behind of those individuals which are in all respects fit for the conditions of their existence; and are therefore very nearly alike. The circumstances to which any species is exposed, being, as we before saw, an involved combination of incident forces; and the members of the species having mixed with them some that differ more than usual from the average structure required for meeting these forces; it results that these forces are constantly separating such divergent individuals from the rest, and so preserving the uniformity of the rest—keeping up its integrity as a species. Just as the changing autumn leaves are picked out by the wind from among the green ones around them, or just as, to use Prof. Huxley's simile, the smaller fragments pass through the sieve while the larger are kept back; so, the uniform incidence of external forces affects the members of a group of organisms similarly in proportion as they are similar, and differently in proportion as they are different; and thus is ever segregating the like by parting the unlike from them. Whether these separated members are killed off, as mostly happens, or whether, as otherwise happens, they survive and multiply into a distinct variety, in consequence of their fitness to certain partially unlike conditions, matters not to the argument. The one case conforms to the law, that the unlike units of an aggregate are differentiated and integrated when uniformly subject to the same incident forces; and the other to the converse law, that the like units of an aggregate are differentiated and integrated when subject to different incident forces. And on consulting Mr. Darwin's remarks on divergence of character, it will be seen that the segregations thus caused tend ever to become more definite.

§ 127. Mental evolution under one of its leading aspects, we found to consist in the formation of groups of like objects and like relations—a differentiation of the various things originally confounded together in one assemblage, and an integration of each separate order of things into a separate group (§ 113). Here it remains to point out that while unlikeness in the incident forces is the cause of such differentiations, likeness in the incident forces is the cause of such integrations. For what is the process through which classifications are established? At first, in common with the uninitiated, the botanist recognizes only such conventional divisions as those which agriculture has established—distinguishes a few vegetables and cereals, and groups the rest together into the one miscellaneous aggregate of wild plants. How do these wild plants become grouped in his mind into orders, genera, and species? Each plant he examines yields him a certain complex impression. Every now and then he picks up a plant like one before seen; and the recognition of it is the production in him of a like connected group of sensations, by a like connected group of attributes. That is to say, there is produced throughout the nerves concerned, a combined set of changes, similar to a combined set of changes before produced. Considered analytically, each such combined set of changes is a combined set of molecular modifications wrought in the affected part of the organism. On every repetition of the impression, a like combined set of molecular modifications is superposed on the previous ones, and makes them greater: thus generating an internal idea corresponding to these similar external objects. Meanwhile, another kind of plant produces in the brain of the botanist another set of combined changes or molecular modifications—a set which does not agree with and deepen the one we have been considering, but disagrees with it; and by repetition of such there is generated a different idea answering to a different species. What now is the nature of this process expressed in general terms? On the one hand there are the like and unlike things from which severally emanate the groups of forces by which we perceive them. On the other hand, there are the organs of sense and percipient centres, through which, in the course of observation, these groups of forces pass. In passing through these organs of sense and percipient centres, the like groups of forces are segregated, or separated from the unlike groups of forces; and each such differentiated and integrated series of groups of forces, answering to an external genus or species, constitutes a state of consciousness which we call our idea of the genus or species. We before saw that as well as a separation of mixed matters by the same force, there is a separation of mixed forces by the same matter; and here we may further see that the unlike forces so separated, work unlike structural changes in the aggregate that separates them—structural changes each of which thus represents, and is equivalent to, the integrated series of motions that has produced it.

By a parallel process, the connexions of co-existence and sequence among impressions, become differentiated and integrated simultaneously with the impressions themselves. When two phenomena that have been experienced in a given order, are repeated in the same order, those nerves which before were affected by the transition are again affected; and such molecular modification as they received from the first motion propagated through them, is increased by this second motion along the same route. Each such motion works a structural alteration, which, in conformity with the general law set forth in Chapter X., involves a diminution of the resistance to all such motions that afterwards occur. The integration of these successive motions (or more strictly, the permanently effective portions of them expended in overcoming resistance) thus becomes the cause of, and the measure of, the mental connexion between the impressions which the phenomena produce. Meanwhile, phenomena that are recognized as different from these, being phenomena that therefore affect different nervous elements, will have their connexions severally represented by motions along other routes; and along each of these other routes, the nervous discharges will severally take place with a readiness proportionate to the frequency with which experience repeats the connexion of phenomena. The classification of relations must hence go on *pari passu* with the classification of the related things. In common with the mixed sensations received from the external world, the mixed relations it presents, cannot be impressed on the organism without more or less segregation of them resulting. And through this continuous differentiation and integration of changes or motions, which constitutes nervous function, there is gradually wrought that differentiation and integration of matter, which constitutes nervous structure.

§ 128. In social evolution, the collecting together of the like and the separation of the unlike, by incident forces, is primarily displayed in the same manner as we saw it to be among groups of inferior creatures. The human races tend to differentiate and integrate, as do races of other living forms. Of the forces which effect and maintain the segregations of mankind, may first be named those external ones which we class as physical conditions. The climate and food that are favourable to an indigenous people, are more or less detrimental to a people of different bodily constitution, coming from a remote part of the Earth. In tropical regions the northern races cannot permanently exist: if not killed off in the first generation, they are so in the second; and, as in India, can maintain their footing only by the artificial process of continuous immigration and emigration. That is to say, the external forces acting equally on the inhabitants of a given locality, tend to expel all who are not of a certain type; and so to keep up the integration of those who are of that type. Though elsewhere, as among European nations, we see a certain amount of permanent intermixture, otherwise brought about, we still see that this takes place between races of not very different types, that are naturalized to not very different conditions. The other forces conspiring to produce these national integrations, are those mental ones which show themselves in the affinities of men for others like themselves. Emigrants usually desire to get back among their own people; and where their desire does not take effect, it is only because the restraining ties are too great. Units of one society who are obliged to reside in another, very generally form colonies in the midst of that other—small societies of their own. Races which have been artificially severed, show strong tendencies to re-unite. Now though these integrations that result from the mutual affinities of kindred men, do not seem interpretable as illustrations of the general principle above enunciated, they really are thus interpretable. When treating of the direction of motion (§ 91), it was shown that the actions performed by men for the satisfaction of their wants, were always motions along lines of least resistance. The feelings characterizing a member of a given race, are feelings which get complete satisfaction only among other members of that race—a satisfaction partly derived from sympathy with those having like feelings, but mainly derived from the adapted social conditions which grow up where such feelings prevail. When, therefore, a citizen of any nation is, as we see, attracted towards others of his nation, the rationale is, that certain agencies which we call desires, move him in the direction of least resistance. Human motions, like all other motions, being determined by the distribution of forces, it follows that such integrations of races as are not produced by incident external forces, are produced by forces which the units of the races exercise on each other.

During the development of each society, we see analogous segregations caused in analogous ways. A few of them result from minor natural affinities; but those most important ones which constitute political and industrial organization, result from the union of men in whom similarities have been produced by education—using education in its widest sense, as comprehending all processes by which citizens are moulded to special functions. Men brought up to bodily labour, are men who have had wrought in them a certain likeness—a likeness which, in respect of their powers of action, obscures and subordinates their natural differences. Those trained to brain-work, have acquired a certain other community of character which makes them, as social units, more like each other than like those trained to manual occupations. And there arise class-integrations answering to these superinduced likenesses. Much more definite integrations take place among the much more definitely assimilated members of any class who are brought up to the same calling. Even where the necessities of their work forbid concentration in one locality, as among artizans happens with masons and brick-layers, and among traders happens with the retail distributors, and among professionals happens with the medical men; there are not wanting Operative Builders Unions, and Grocers Societies, and Medical Associations, to show that these artificially-assimilated citizens become integrated as much as the conditions permit. And where, as among the manufacturing classes, the functions discharged do not require the dispersion of the citizens thus artificially assimilated, there is a progressive aggregation of them in special localities; and a consequent increase in the definiteness of the industrial divisions. If now we seek the causes of these integrations, considered as results of force and motion, we find ourselves brought to the same general principle as before. This likeness generated in any class or subclass by training, is an aptitude acquired by its members for satisfying their wants in like ways. That is, the occupation to which each man has been brought up, has become to him, in common with those similarly brought up, a line of least resistance. Hence under that pressure which determines all men to activity, these similarly-modified social units are similarly affected, and tend to take similar courses. If then there be any locality which, either by its physical peculiarities or by peculiarities wrought on it during social evolution, is rendered a place where a certain kind of industrial action meets with less resistance than elsewhere; it follows from the law of direction of motion that those social units who have been moulded to this kind of industrial action, will move towards this place, or become integrated there. If, for instance, the proximity of coal and iron mines to a navigable river, gives to Glasgow a certain advantage in the building of iron ships—if the total labour required to produce the same vessel, and get its equivalent in food and clothing, is less there than elsewhere; a concentration of iron-ship builders is produced at Glasgow: either by keeping there the population born to iron-ship building; or by immigration of those elsewhere engaged in it; or by both—a concentration that would be still more marked did not other districts offer counter-balancing facilities. The principle equally holds where the occupation is mercantile instead of manufacturing. Stock-brokers cluster together in the city, because the amount of effort to be severally gone through by them in discharging their functions, and obtaining their profits, is less there than in other localities. A place of exchange having once been established, becomes a place where the resistance to be overcome by each is less than elsewhere; and the pursuit of the course of least resistance by each, involves their aggregation around this place.

Of course, with units so complicated as those which constitute a society, and with forces so involved as those which move them, the resulting differentiations and integrations must be far more entangled, or far less definite, than those we have hitherto considered. But though there may be pointed out many anomalies which at first sight seem inconsistent with the alleged law, a closer study shows that they are but subtler illustrations of it. For men's likenesses being of various kinds, lead to various order of integration. There are likenesses of disposition, likenesses of taste, likenesses produced by intellectual culture, likenesses that result from class-training, likenesses of political feeling; and it needs but to glance round at the caste-divisions, the associations for philanthropic, scientific, and artistic purposes, the religious parties and social cliques; to see that some species of likeness among the component members of each body determines their union. Now these different integrations, by traversing each other, and often by their indirect antagonism, more or less obscure each other; and prevent any one kind of integration from becoming complete. Hence the anomalies referred to. But if this cause of incompleteness be duly borne in mind, social segregations will be seen to conform entirely to the same principle as all other segregations. Analysis will show that either by external incident forces, or by what we may in a sense regard as mutual polarity, there are ever being produced in society integrations of those units which have either a natural likeness or a likeness generated by training.

§ 129. Can the general truth thus variously illustrated be deduced from the persistence of force, in common with foregoing ones? Probably the exposition at the beginning of the chapter will have led most readers to conclude that it can be so deduced.

The abstract propositions involved are these:—First, that like units, subject to a uniform force capable of producing motion in them, will be moved to like degrees in the same direction. Second, that like units if exposed to unlike forces capable of producing motion in them, will be differently moved—moved either in different directions or to different degrees in the same direction. Third, that unlike units if acted on by a uniform force capable of producing motion in them, will be differently moved—moved either in different directions or to different degrees in the same direction. Fourth, that the incident forces themselves must be affected in analogous ways: like forces falling on like units must be similarly modified by the conflict; unlike forces falling on like units must be dissimilarly modified; and like forces falling on unlike units must be dissimilarly modified. These propositions admit of reduction to a still more abstract form. They all of them amount to this:—that in the actions and reactions of force and matter, an unlikeness in either of the factors necessitates an unlikeness in the effects; and that in the absence of unlikeness in either of the factors the effects must be alike.

When thus generalized, the immediate dependence of these propositions on the persistence of force, becomes obvious. Any two forces that are not alike, are forces which differ either in their amounts or directions or both; and by what mathematicians call the resolution of forces, it may be proved that this difference is constituted by the presence in the one of some force not present in the other. Similarly, any two units or portions of matter which are unlike in size, weight, form, or other attribute, can be known by us as unlike only through some unlikeness in the forces they impress on our consciousness; and hence this unlikeness also, is constituted by the presence in the one of some force or forces not present in the other. Such being the common nature of these unlikenesses, what is the inevitable corollary? Any unlikeness in the incident forces, where the things acted on are alike, must generate a difference between the effects; since otherwise, the differential force produces no effect, and force is not persistent. Any unlikeness in the things acted on, where the incident forces are alike, must generate a difference between the effects; since otherwise, the differential force whereby these things are made unlike, produces no effect, and force is not persistent. While, conversely, if the forces acting and the things acted on, are alike, the effects must be alike; since otherwise, a differential effect can be produced without a differential cause, and force is not persistent.

Thus these general truths being necessary implications of the persistence of force, all the re-distributions above traced out as characterizing Evolution in its various phases, are also implications of the persistence of force. Such portions of the permanently effective forces acting on any aggregate, as produce sensible motions in its parts, cannot but work the segregations which we see take place. If of the mixed units making up such aggregate, those of the same kind have like motions impressed on them by a uniform force, while units of another kind are moved by this uniform force in ways more or less unlike the ways in which those of the first kind are moved, the two kinds must separate and integrate. If the units are alike and the forces unlike, a division of the differently affected units is equally necessitated. Thus there inevitably arises the demarcated grouping which we everywhere see. By virtue of this segregation that grows ever more decided while there remains any possibility of increasing it, the change from uniformity to multiformity is accompanied by a change from indistinctness in the relations of parts to distinctness in the relations of parts. As we before saw that the transformation of the homogeneous into the heterogeneous is inferrable from that ultimate truth which transcends proof; so we here see, that from this same truth is inferrable the transformation of an indefinite homogeneity into a definite heterogeneity.

CHAPTER XVI
EQUILIBRATION

§ 130. And now towards what do these changes tend? Will they go on for ever? or will there be an end to them? Can things increase in heterogeneity through all future time? or must there be a degree which the differentiation and integration of Matter and Motion cannot pass? Is it possible for this universal metamorphosis to proceed in the same general course indefinitely? or does it work towards some ultimate state, admitting no further modification of like kind? The last of these alternative conclusions is that to which we are inevitably driven. Whether we watch concrete processes, or whether we consider the question in the abstract, we are alike taught that Evolution has an impassable limit.

The re-distributions of matter that go on around us, are ever being brought to conclusions by the dissipation of the motions which effect them. The rolling stone parts with portions of its momentum to the things it strikes, and finally comes to rest; as do also, in like manner, the various things it has struck. Descending from the clouds and trickling over the Earth's surface till it gathers into brooks and rivers, water, still running towards a lower level, is at last arrested by the resistance of other water that has reached the lowest level. In the lake or sea thus formed, every agitation raised by a wind or the immersion of a solid body, propagates itself around in waves that diminish as they widen, and gradually become lost to observation in motions communicated to the atmosphere and the matter on the shores. The impulse given by a player to the harp-string, is transformed through its vibrations into aerial pulses; and these, spreading on all sides, and weakening as they spread, soon cease to be perceptible; and finally die away in generating thermal undulations that radiate into space. Equally in the cinder that falls out of the fire, and in the vast masses of molten lava ejected by a volcano, we see that the molecular agitation known to us as heat, disperses itself by radiation; so that however great its amount, it inevitably sinks at last to the same degree as that existing in surrounding bodies. And if the actions observed be electrical or chemical, we still find that they work themselves out in producing sensible or insensible movements, that are dissipated as before; until quiescence is eventually reached. The proximate rationale of the process exhibited under these several forms, lies in the fact dwelt on when treating of the Multiplication of Effects, that motions are ever being decomposed into divergent motions, and these into re-divergent motions. The rolling stone sends off the stones it hits in directions differing more or less from its own; and they do the like with the things they hit. Move water or air, and the movement is quickly resolved into radiating movements. The heat produced by pressure in a given direction, diffuses itself by undulations in all directions; and so do the light and electricity similarly generated. That is to say, these motions undergo division and subdivision; and by continuance of this process without limit, they are, though never lost, gradually reduced to insensible motions.

In all cases then, there is a progress toward equilibration. That universal co-existence of antagonist forces which, as we before saw, necessitates the universality of rhythm, and which, as we before saw, necessitates the decomposition of every force into divergent forces, at the same time necessitates the ultimate establishment of a balance. Every motion being motion under resistance, is continually suffering deductions; and these unceasing deductions finally result in the cessation of the motion.

The general truth thus illustrated under its simplest aspect, we must now look at under those more complex aspects it usually presents throughout Nature. In nearly all cases, the motion of an aggregate is compound; and the equilibration of each of its components, being carried on independently, does not affect the rest. The ship's bell that has ceased to vibrate, still continues those vertical and lateral oscillations caused by the ocean-swell. The water of the smooth stream on whose surface have died away the undulations caused by the rising fish, moves as fast as before onward to the sea. The arrested bullet travels with undiminished speed round the Earth's axis. And were the rotation of the Earth destroyed, there would not be implied any diminution of the Earth's movement with respect to the Sun and other external bodies. So that in every case, what we regard as equilibration is a disappearance of some one or more of the many movements which a body possesses, while its other movements continue as before. That this process may be duly realized and the state of things towards which it tends fully understood, it will be well here to cite a case in which we may watch this successive equilibration of combined movements more completely than we can do in those above instanced. Our end will best be served, not by the most imposing, but by the most familiar example. Let us take that of the spinning top. When the string which has been wrapped round a top's axis is violently drawn off, and the top falls on to the table, it usually happens that besides the rapid rotation, two other movements are given to it. A slight horizontal momentum, unavoidably impressed on it when leaving the handle, carries it away bodily from the place on which it drops; and in consequence of its axis being more or less inclined, it falls into a certain oscillation, described by the expressive though inelegant word—"wabbling." These two subordinate motions, variable in their proportions to each other and to the chief motion, are commonly soon brought to a close by separate processes of equilibration. The momentum which carries the top bodily along the table, resisted somewhat by the air, but mainly by the irregularities of the surface, shortly disappears; and the top thereafter continues to spin on one spot. Meanwhile, in consequence of that opposition which the axial momentum of a rotating body makes to any change in the plane of rotation, (so beautifully exhibited by the gyroscope,) the "wabbling" diminishes; and like the other is quickly ended. These minor motions having been dissipated, the rotatory motion, interfered with only by atmospheric resistance and the friction of the pivot, continues some time with such uniformity that the top appears stationary: there being thus temporarily established a condition which the French mathematicians have termed *equilibrium mobile*. It is true that when the axial velocity sinks below a certain point, new motions commence, and increase till the top falls; but these are merely incidental to a case in which the centre of gravity is above the point of support. Were the top, having an axis of steel, to be suspended from a surface adequately magnetized, all the phenomena described would be displayed, and the moving equilibrium having been once arrived at, would continue until the top became motionless, without any further change of position. Now the facts which it behoves us here to observe, are these. First, that the various motions which an aggregate possesses are separately equilibrated: those which are smallest, or which meet with the greatest resistance, or both, disappearing first; and leaving at last, that which is greatest, or meets with least resistance, or both. Second, that when the aggregate has a movement of its parts with respect to each other, which encounters but little external resistance, there is apt to be established an *equilibrium mobile*. Third, that this moving equilibrium eventually lapses into complete equilibrium.

Fully to comprehend the process of equilibration, is not easy; since we have simultaneously to contemplate various phases of it. The best course will be to glance separately at what we may conveniently regard as its four different orders. The first order includes the comparatively simple motions, as those of projectiles, which are not prolonged enough to exhibit their rhythmical character; but which, being quickly divided and subdivided into motions communicated to other portions of matter, are presently dissipated in the rhythm of ethereal undulations. In the second order, comprehending the various kinds of vibration or oscillation as usually witnessed, the motion is used up in generating a tension which, having become equal to it or momentarily equilibrated with it, thereupon produces a motion in the opposite direction, that is subsequently equilibrated in like manner: thus causing a visible rhythm, that is, however, soon lost in invisible rhythms. The third order of equilibration, not hitherto noticed, obtains in those aggregates which continually receive as much motion as they expend. The steam engine (and especially that kind which feeds its own furnace and boiler) supplies an example. Here the force from moment to moment dissipated in overcoming the resistance of the machinery driven, is from moment to moment replaced from the fuel; and the balance of the two is maintained by a raising or lowering of the expenditure according to the variation of the supply: each increase or decrease in the quantity of steam, resulting in a rise or fall of the engine's movement, such as brings it to a balance with the increased or decreased resistance. This, which we may fitly call the *dependent* moving equilibrium, should be specially noted; since it is one that we shall commonly meet with throughout various phases of Evolution. The equilibration to be distinguished as of the fourth order, is the *independent* or perfect moving equilibrium. This we see illustrated in the rhythmical motions of the Solar System; which, being resisted only by a medium of inappreciable density, undergo no sensible diminution in such periods of time as we can measure.

All these kinds of equilibration may, however, from the highest point of view, be regarded as different modes of one kind. For in every case the balance arrived at is relative, and not absolute—is a cessation of the motion of some particular body in relation to a certain point or points, involving neither the disappearance of the relative motion lost, which is simply transformed into other motions, nor a diminution of the body's motions with respect to other points. Thus understanding equilibration, it manifestly includes that *equilibrium mobile*, which at first sight seems of another nature. For any system of bodies exhibiting, like those of the Solar System, a combination of balanced rhythms, has this peculiarity;—that though the constituents of the system have relative movements, the system as a whole has no movement. The centre of gravity of the entire group remains fixed. Whatever quantity of motion any member of it has in any direction, is from moment to moment counter-balanced by an equivalent motion in some other part of the group in an opposite direction; and so the aggregate matter of the group is in a state of rest. Whence it follows that the arrival at a state of moving equilibrium, is the disappearance of some movement which the aggregate had in relation to external things, and a continuance of those movements only which the different parts of the aggregate have in relation to each other. Thus generalizing the process, it becomes clear that all forms of equilibration are intrinsically the same; since in every aggregate, it is the centre of gravity only that loses its motion: the constituents always retaining some motion with respect to each other—the motion of molecules if none else.

Those readers who happen to bear in mind a proposition concerning the functional characteristics of Evolution, which was set forth in Chapter XII, will probably regard it as wholly at variance with that set forth in this Chapter. It was there alleged that throughout Evolution, integration of matter is accompanied by integration of such motion as the matter previously had; and that thus there is a transformation of diffused motion into aggregated motion, parallel to the transformation of diffused matter into aggregated matter. Here however, it is asserted that every aggregate motion is constantly undergoing diffusion—every integrated motion undergoing perpetual disintegration. And so the motion of masses, which before was said gradually to arise out of molecular motion, is here said to be gradually lost in molecular motion. Doubtless these statements, if severally accepted without qualification, are contradictory. Neither of them, however, expresses the whole truth. Each needs the other as its indispensable complement. It is quite true, as before alleged, that there goes on an integration of motion corresponding to the integration of matter; and that this essential characteristic of Evolution, functionally considered, is clearly displayed in proportion as the Evolution is active. But the disintegration of motion, which, as we before saw, constitutes Dissolution, functionally considered, is all along going on; and though at first it forms but a small deduction from the change constituting Evolution, it gradually becomes equal to it, and eventually exceeding it, entails reverse changes. The aggregation of matter never being complete, but leaving behind less aggregated or unaggregated matter, in the shape of liquid, aeriform, or ethereal media; it results that from the beginning, the integrated motion of integrated masses, is ever being obstructed by these less integrated or unintegrated media. So that though while the integration of matter is rapidly going on, there is an increase of integrated motion, spite of the deductions thus continually made from it, there comes a time when the integration of matter and consequently of motion, ceases to increase, or increases so slowly that the deductions counterbalance it; and thenceforth these begin to decrease it, and, by its perpetual diffusion, to bring about a relative equilibration. From the beginning, the process of Evolution is antagonized by a process of Dissolution; and while the first for a long time predominates, the last finally arrests and reverses it.

Returning from this parenthetical explanation, we must now especially note two leading truths brought out by the foregoing exposition: the one concerning the ultimate, or rather the penultimate, state of motion which the processes described tend to bring about; the other concerning the concomitant distribution of matter. This penultimate state of motion is the moving equilibrium; which, as we have seen, tends to arise in an aggregate having compound motions, as a transitional state on the way towards complete equilibrium. Throughout Evolution of all kinds, there is a continual approximation to, and more or less complete maintenance of, this moving equilibrium. As in the Solar System there has been established an independent moving equilibrium—an equilibrium such that the relative motions of the constituent parts are continually so counter-balanced by opposite motions, that the mean state of the whole aggregate never varies; so is it, though in a less distinct manner, with each form of dependent moving equilibrium. The state of things exhibited in the cycles of terrestrial changes, in the balanced functions of organic bodies that have reached their adult forms, and in the acting and re-acting processes of fully-developed societies, is similarly one characterized by compensating oscillations. The involved combination of rhythms seen in each of these cases, has an average condition which remains practically constant during the deviations ever taking place on opposite sides of it. And the fact which we have here particularly to observe, is, that as a corollary from the general law of equilibration above set forth, the evolution of every aggregate must go on until this *equilibrium mobile* is established; since, as we have seen, an excess of force which the aggregate possesses in any direction, must eventually be expended in overcoming resistances to change in that direction: leaving behind only those movements which compensate each other, and so form a moving equilibrium. Respecting the structural state simultaneously reached, it must obviously be one presenting an arrangement of forces that counterbalance all the forces to which the aggregate is subject. So long as there remains a residual force in any direction—be it excess of a force exercised by the aggregate on its environment, or of a force exercised by its environment on the aggregate, equilibrium does not exist; and therefore the re-distribution of matter must continue. Whence it follows that the limit of heterogeneity towards which every aggregate progresses, is the formation of as many specializations and combinations of parts, as there are specialized and combined forces to be met.

§ 131. Those successively changed forms which, if the nebular hypothesis be granted, must have arisen during the evolution of the Solar System, were so many transitional kinds of moving equilibrium; severally giving place to more permanent kinds on the way towards complete equilibration. Thus the assumption of an oblate spheroidal figure by condensing nebulous matter, was the assumption of a temporary and partial moving equilibrium among the component parts—a moving equilibrium that must have slowly grown more settled, as local conflicting movements were dissipated. In the formation and detachment of the nebulous rings, which, according to this hypothesis, from time to time took place, we have instances of progressive equilibration ending in the establishment of a complete moving equilibrium. For the genesis of each such ring, implies a perfect balancing of that aggregative force which the whole spheroid exercises on its equatorial portion, by that centrifugal force which the equatorial portion has acquired during previous concentration: so long as these two forces are not equal, the equatorial portion follows the contracting mass; but as soon as the second force has increased up to an equality with the first, the equatorial portion can follow no further, and remains behind. While, however, the resulting ring, regarded as a whole connected by forces with external wholes, has reached a state of moving equilibrium; its parts are not balanced with respect to each other. As we before saw (§ 110) the probabilities against the maintenance of an annular form by nebulous matter, are immense: from the instability of the homogeneous, it is inferrable that nebulous matter so distributed must break up into portions; and eventually concentrate into a single mass. That is to say, the ring must progress towards a moving equilibrium of a more complete kind, during the dissipation of that motion which maintained its particles in a diffused form: leaving at length a planetary body, attended perhaps by a group of minor bodies, severally having residuary relative motions that are no longer resisted by sensible media; and there is thus constituted an *equilibrium mobile* that is all but absolutely perfect.[18]

Hypothesis aside, the principle of equilibration is still perpetually illustrated in those minor changes of state which the Solar System is undergoing. Each planet, satellite, and comet, exhibits to us at its aphelion a momentary equilibrium between that force which urges it further away from its primary, and that force which retards its retreat; since the retreat goes on until the last of these forces exactly counterpoises the first. In like manner at perihelion a converse equilibrium is momentarily established. The variation of each orbit in size, in eccentricity, and in the position of its plane, has similarly a limit at which the forces producing change in the one direction, are equalled by those antagonizing it; and an opposite limit at which an opposite arrest takes place. Meanwhile, each of these simple perturbations, as well as each of the complex ones resulting from their combination, exhibits, besides the temporary equilibration at each of its extremes, a certain general equilibration of compensating deviations on either side of a mean state. That the moving equilibrium thus constituted, tends, in the course of indefinite time, to lapse into a complete equilibrium, by the gradual decrease of planetary motions and eventually integration of all the separate masses composing the Solar System, is a belief suggested by certain observed cometary retardations, and entertained by some of high authority. The received opinion that the appreciable diminution in the period of Encke's comet, implies a loss of momentum caused by resistance of the ethereal medium, commits astronomers who hold it, to the conclusion that this same resistance must cause a loss of planetary motions—a loss which, infinitesimal though it may be in such periods as we can measure, will, if indefinitely continued, bring these motions to a close. Even should there be, as Sir John Herschel suggests, a rotation of the ethereal medium in the same direction with the planets, this arrest, though immensely postponed, would not be absolutely prevented. Such an eventuality, however, must in any case be so inconceivably remote as to have no other than a speculative interest for us. It is referred to here, simply as illustrating the still-continued tendency towards complete equilibrium, through the still-continued dissipation of sensible motion, or transformation of it into insensible motion.

But there is another species of equilibration going on in the Solar System, with which we are more nearly concerned—the equilibration of that molecular motion known as heat. The tacit assumption hitherto current, that the Sun can continue to give off an undiminished amount of light and heat through all future time, is fast being abandoned. Involving as it does, under a disguise, the conception of power produced out of nothing, it is of the same order as the belief that misleads perpetual-motion schemers. The spreading recognition of the truth that force is persistent, and that consequently whatever force is manifested under one shape must previously have existed under another shape, is carrying with it a recognition of the truth that the force known to us in solar radiations, is the changed form of some other force of which the Sun is the seat; and that by the gradual dissipation of these radiations into space, this other force is being slowly exhausted. The aggregative force by which the Sun's substance is drawn to his centre of gravity, is the only one which established physical laws warrant us in suspecting to be the correlate of the forces thus emanating from him: the only source of a known kind that can be assigned for the insensible motions constituting solar light and heat, is the sensible motion which disappears during the progressing concentration of the Sun's substance. We before saw it to be a corollary from the nebular hypothesis, that there is such a progressing concentration of the Sun's substance. And here remains to be added the further corollary, that just as in the case of the smaller members of the Solar System, the heat generated by concentration, long ago in great part radiated into space, has left only a central residue that now escapes but slowly; so in the case of that immensely larger mass forming the Sun, the immensely greater quantity of heat generated and still in process of rapid diffusion, must, as the concentration approaches its limit, diminish in amount, and eventually leave only an inappreciable internal remnant. With or without the accompaniment of that hypothesis of nebular condensation, whence, as we see, it naturally follows, the doctrine that the Sun is gradually losing his heat, has now gained considerable currency; and calculations have been made, both respecting the amount of heat and light already radiated, as compared with the amount that remains, and respecting the period during which active radiation is likely to continue. Prof. Helmholtz estimates, that since the time when, according to the nebular hypothesis, the matter composing the Solar System extended to the orbit of Neptune, there has been evolved by the arrest of sensible motion, an amount of heat 454 times as great as that which the Sun still has to give out. He also makes an approximate estimate of the rate at which this remaining 1 454th is being diffused: showing that a diminution of the Sun's diameter to the extent of 1 10,000, would produce heat, at the present rate, for more than 2000 years; or in other words, that a contraction of 1 20,000,000 of his diameter, suffices to generate the amount of light and heat annually emitted; and that thus, at the present rate of expenditure, the Sun's diameter will diminish by something like 1 20 in the lapse of the next million years.[19] Of course these conclusions are not to be considered as more than rude approximations to the truth. Until quite recently, we have been totally ignorant of the Sun's chemical composition; and even now have obtained but a superficial knowledge of it. We know nothing of his internal structure; and it is quite possible (probable, I believe,) that the assumptions respecting central density, made in the foregoing estimates, are wrong. But no uncertainty in the data on which these calculations proceed, and no consequent error in the inferred rate at which the Sun is expending his reserve of force, militates against the general proposition that this reserve of force *is* being expended; and must in time be exhausted. Though the residue of undiffused motion in the Sun, may be much greater than is above concluded; though the rate of radiation cannot, as assumed, continue at a uniform rate, but must eventually go on with slowly-decreasing rapidity; and though the period at which the Sun will cease to afford us adequate light and heat, is very possibly far more distant than above implied; yet such a period must some time be reached, and this is all which it here concerns us to observe.

Thus while the Solar System, if evolved from diffused matter, has illustrated the law of equilibration in the establishment of a complete moving equilibrium; and while, as at present constituted, it illustrates the law of equilibration in the balancing of all its movements; it also illustrates this law in the processes which astronomers and physicists infer are still going on. That motion of masses produced during Evolution, is being slowly re-diffused in molecular motion of the ethereal medium; both through the progressive integration of each mass, and the resistance to its motion through space. Infinitely remote as may be the state when all the motions of masses shall be transformed into molecular motion, and all the molecular motion equilibrated; yet such a state of complete integration and complete equilibration, is that towards which the changes now going on throughout the Solar System inevitably tend.

§ 132. A spherical figure is the one which can alone equilibrate the forces of mutually-gravitating atoms. If the aggregate of such atoms has a rotatory motion, the form of equilibrium becomes a spheroid of greater or less oblateness, according to the rate of rotation; and it has been ascertained that the Earth is an oblate spheroid, diverging just as much from sphericity as is requisite to counterbalance the centrifugal force consequent on its velocity round its axis. That is to say, during the evolution of the Earth, there has been reached a complete equilibrium of those forces which affect its general outline. The only other process of equilibration which the Earth as a whole can exhibit, is the loss of its axial motion; and that any such loss is going on, we have no direct evidence. It has been contended, however, by Prof. Helmholtz, that inappreciable as may be its effect within known periods of time, the friction of the tidal wave must be slowly diminishing the Earth's rotatory motion, and must eventually destroy it. Now though it seems an oversight to say that the Earth's rotation can thus be destroyed, since the extreme effect, to be reached only in infinite time by such a process, would be an extension of the Earth's day to the length of a lunation; yet it seems clear that this friction of the tidal wave is a real cause of decreasing rotation. Slow as its action is, we must recognize it as exemplifying, under another form, the universal progress towards equilibrium.

It is needless to point out, in detail, how those movements which the Sun's rays generate in the air and water on the Earth's surface, and through them in the Earth's solid substance,[20] one and all teach the same general truth. Evidently the winds and waves and streams, as well as the denudations and depositions they effect, perpetually illustrate on a grand scale, and in endless modes, that gradual dissipation of motions described in the first section; and the consequent tendency towards a balanced distribution of forces. Each of these sensible motions, produced directly or indirectly by integration of those insensible motions communicated from the Sun, becomes, as we have seen, divided and subdivided into motions less and less sensible; until it is finally reduced to insensible motions, and radiated from the Earth in the shape of thermal undulations. In their totality, these complex movements of aerial, liquid, and solid matter on the Earth's crust, constitute a dependent moving equilibrium. As we before saw, there is traceable throughout them an involved combination of rhythms. The unceasing circulation of water from the ocean to the land, and from the land back to the ocean, is a type of these various compensating actions; which, in the midst of all the irregularities produced by their mutual interferences, maintain an average. And in this, as in other equilibrations of the third order, we see that the power from moment to moment in course of dissipation, is from moment to moment renewed from without: the rises and falls in the supply, being balanced by rises and falls in the expenditure; as witness the correspondence between the magnetic variations and the cycle of the solar spots. But the fact it chiefly concerns us to observe, is, that this process must go on bringing things ever nearer to complete rest. These mechanical movements, meteorologic and geologic, which are continually being equilibrated, both temporarily by counter-movements and permanently by the dissipation of such movements and counter-movements, will slowly diminish as the quantity of force received from the Sun diminishes. As the insensible motions propagated to us from the centre of our system become feebler, the sensible motions here produced by them must decrease; and at that remote period when the solar heat has ceased to be appreciable, there will no longer be any appreciable re-distributions of matter on the surface of our planet.

Thus from the highest point of view, all terrestrial changes are incidents in the course of cosmical equilibration. It was before pointed out, (§ 80) that of the incessant alterations which the Earth's crust and atmosphere undergo, those which are not due to the still-progressing motion of the Earth's substance towards its centre of gravity, are due to the still-progressing motion of the Sun's substance towards its centre of gravity. Here it is to be remarked, that this continuance of integration in the Earth and in the Sun, is a continuance of that transformation of sensible motion into insensible motion which we have seen ends in equilibration; and that the arrival in each case at the extreme of integration, is the arrival at a state in which no more sensible motion remains to be transformed into insensible motion—a state in which the forces producing integration and the forces opposing integration, have become equal.

§ 133. Every living body exhibits, in a four-fold form, the process we are tracing out—exhibits it from moment to moment in the balancing of mechanical forces; from hour to hour in the balancing of functions; from year to year in the changes of state that compensate changes of condition; and finally in the complete arrest of vital movements at death. Let us consider the facts under these heads.

The sensible motion constituting each visible action of an organism, is soon brought to a close by some adverse force within or without the organism. When the arm is raised, the motion given to it is antagonized partly by gravity and partly by the internal resistances consequent on structure; and its motion, thus suffering continual deduction, ends when the arm has reached a position at which the forces are equilibrated. The limits of each systole and diastole of the heart, severally show us a momentary equilibrium between muscular strains that produce opposite movements; and each gush of blood requires to be immediately followed by another, because the rapid dissipation of its momentum would otherwise soon bring the mass of circulating fluid to a stand. As much in the actions and re-actions going on among the internal organs, as in the mechanical balancing of the whole body, there is at every instant a progressive equilibration of the motions at every instant produced. Viewed in their aggregate, and as forming a series, the organic functions constitute a dependent moving equilibrium—a moving equilibrium, of which the motive power is ever being dissipated through the special equilibrations just exemplified, and is ever being renewed by the taking in of additional motive power. Food is a store of force which continually adds to the momentum of the vital actions, as much as is continually deducted from them by the forces overcome. All the functional movements thus maintained, are, as we have seen, rhythmical (§ 96); by their union compound rhythms of various lengths and complexities are produced; and in these simple and compound rhythms, the process of equilibration, besides being exemplified at each extreme of every rhythm, is seen in the habitual preservation of a constant mean, and in the re-establishment of that mean when accidental causes have produced divergence from it. When, for instance, there is a great expenditure of motion through muscular activity, there arises a re-active demand on those stores of latent motion which are laid up in the form of consumable matter throughout the tissues: increased respiration and increased rapidity of circulation, are instrumental to an extra genesis of force, that counter-balances the extra dissipation of force. This unusual transformation of molecular motion into sensible motion, is presently followed by an unusual absorption of food—the source of molecular motion; and in proportion as there has been a prolonged draft upon the spare capital of the system, is there a tendency to a prolonged rest, during which that spare capital is replaced. If the deviation from the ordinary course of the functions has been so great as to derange them, as when violent exertion produces loss of appetite and loss of sleep, an equilibration is still eventually effected. Providing the disturbance is not such as to overturn the balance of the functions, and destroy life (in which case a complete equilibration is suddenly effected), the ordinary balance is by and by re-established: the returning appetite is keen in proportion as the waste has been large; while sleep, sound and prolonged, makes up for previous wakefulness. Not even in those extreme cases where some excess has wrought a derangement that is never wholly rectified, is there an exception to the general law; for in such cases the cycle of the functions is, after a time, equilibrated about a new mean state, which thenceforth becomes the normal state of the individual. Thus, among the involved rhythmical changes constituting organic life, any disturbing force that works an excess of change in some direction, is gradually diminished and finally neutralized by antagonistic forces; which thereupon work a compensating change in the opposite direction, and so, after more or less of oscillation, restore the medium condition. And this process it is, which constitutes what physicians call the *vis medicatrix naturæ*. The third form of equilibration displayed by organic bodies, is a necessary sequence of that just illustrated. When through a change of habit or circumstance, an organism is permanently subject to some new influence, or different amount of an old influence, there arises, after more or less disturbance of the organic rhythms, a balancing of them around the new average condition produced by this additional influence. As temporary divergences of the organic rhythms are counteracted by temporary divergences of a reverse kind; so there is an equilibration of their permanent divergences by the genesis of opposing divergences that are equally permanent. If the quantity of motion to be habitually generated by a muscle, becomes greater than before, its nutrition becomes greater than before. If the expenditure of the muscle bears to its nutrition, a greater ratio than expenditure bears to nutrition in other parts of the system; the excess of nutrition becomes such that the muscle grows. And the cessation of its growth is the establishment of a balance between the daily waste and the daily repair—the daily expenditure of force, and the amount of latent force daily added. The like must manifestly be the case with all organic modifications consequent on change of climate or food. This is a conclusion which we may safely draw without knowing the special re-arrangements that effect the equilibration. If we see that a different mode of life is followed, after a period of functional derangement, by some altered condition of the system—if we see that this altered condition, becoming by and by established, continues without further change; we have no alternative but to say, that the new forces brought to bear on the system, have been compensated by the opposing forces they have evoked. And this is the interpretation of the process which we call *adaptation*. Finally, each organism illustrates the law in the *ensemble* of its life. At the outset it daily absorbs under the form of food, an amount of force greater than it daily expends; and the surplus is daily equilibrated by growth. As maturity is approached, this surplus diminishes; and in the perfect organism, the day's absorption of potential motion balances the day's expenditure of actual motion. That is to say, during adult life, there is continuously exhibited an equilibration of the third order. Eventually, the daily loss, beginning to out-balance the daily gain, there results a diminishing amount of functional action; the organic rhythms extend less and less widely on each side of the medium state; and there finally results that complete equilibration which we call death.

The ultimate structural state accompanying that ultimate functional state towards which an organism tends, both individually and as a species, may be deduced from one of the propositions set down in the opening section of this chapter. We saw that the limit of heterogeneity is arrived at whenever the equilibration of any aggregate becomes complete—that the re-distribution of matter can continue so long only as there continues any motion unbalanced. Whence we found it to follow that the final structural arrangements, must be such as will meet all the forces acting on the aggregate, by equivalent antagonist forces. What is the implication in the case of organic aggregates; the equilibrium of which is a moving one? We have seen that the maintenance of such a moving equilibrium, requires the habitual genesis of internal forces corresponding in number, directions, and amounts to the external incident forces—as many inner functions, single or combined, as there are single or combined outer actions to be met. But functions are the correlatives of organs; amounts of functions are, other things equal, the correlatives of sizes of organs; and combinations of functions the correlatives of connections of organs. Hence the structural complexity accompanying functional equilibration, is definable as one in which there are as many specialized parts as are capable, separately and jointly, of counteracting the separate and joint forces amid which the organism exists. And this is the limit of organic heterogeneity; to which man has approached more nearly than any other creature.

Groups of organisms display this universal tendency towards a balance very obviously. In § 96, every species of plant and animal was shown to be perpetually undergoing a rhythmical variation in number—now from abundance of food and absence of enemies rising above its average; and then by a consequent scarcity of food and abundance of enemies being depressed below its average. And here we have to observe that there is thus maintained an equilibrium between the sum of those forces which result in the increase of each race, and the sum of those forces which result in its decrease. Either limit of variation is a point at which the one set of forces, before in excess of the other, is counterbalanced by it. And amid these oscillations produced by their conflict, lies that average number of the species at which its expansive tendency is in equilibrium with surrounding repressive tendencies. Nor can it be questioned that this balancing of the preservative and destructive forces which we see going on in every race, must necessarily go on. Since increase of number cannot but continue until increase of mortality stops it; and decrease of number cannot but continue until it is either arrested by fertility or extinguishes the race entirely.

§ 134. The equilibrations of those nervous actions which constitute what we know as mental life, may be classified in like manner with those which constitute what we distinguish as bodily life. We may deal with them in the same order.

Each pulse of nervous force from moment to moment generated, (and it was shown in § 97 that nervous currents are not continuous but rhythmical) is met by counteracting forces; in overcoming which it is dispersed and equilibrated. When tracing out the correlation and equivalence of forces, we saw that each sensation and emotion, or rather such part of it as remains after the excitation of associated ideas and feelings, is expended in working bodily changes—contractions of the involuntary muscles, the voluntary muscles, or both; as also in a certain stimulation of secreting organs. That the movements thus initiated are ever being brought to a close by the opposing forces they evoke, was pointed out above; and here it is to be observed that the like holds with the nervous changes thus initiated. Various facts prove that the arousing of a thought or feeling, always involves the overcoming of a certain resistance: instance the fact that where the association of mental states has not been frequent, a sensible effort is needed to call up the one after the other; instance the fact that during nervous prostration there is a comparative inability to think—the ideas will not follow one another with the habitual rapidity; instance the converse fact that at times of unusual energy, natural or artificial, the friction of thought becomes relatively small, and more numerous, more remote, or more difficult connections of ideas are formed. That is to say, the wave of nervous energy each instant generated, propagates itself throughout body and brain, along those channels which the conditions at the instant render lines of least resistance; and spreading widely in proportion to its amount, ends only when it is equilibrated by the resistances it everywhere meets. If we contemplate mental actions us extending over hours and days, we discover equilibrations analogous to those hourly and daily established among the bodily functions. In the one case as in the other, there are rhythms which exhibit a balancing of opposing forces at each extreme, and the maintenance of a certain general balance. This is seen in the daily alternation of mental activity and mental rest—the forces expended during the one being compensated by the forces acquired during the other. It is also seen in the recurring rise and fall of each desire: each desire reaching a certain intensity, is equilibrated either by expenditure of the force it embodies, in the desired actions, or, less completely, in the imagination of such actions: the process ending in that satiety, or that comparative quiescence, forming the opposite limit of the rhythm. And it is further manifest under a two-fold form, on occasions of intense joy or grief: each paroxysm of passion, expressing itself in vehement bodily actions, presently reaches an extreme whence the counteracting forces produce a return to a condition of moderate excitement; and the successive paroxysms finally diminishing in intensity, end in a mental equilibrium either like that before existing, or partially differing from it in its medium state. But the species of mental equilibration to be more especially noted, is that shown in the establishment of a correspondence between relations among our states of consciousness and relations in the external world. Each outer connection of phenomena which we are capable of perceiving, generates, through accumulated experiences, an inner connection of mental states; and the result towards which this process tends, is the formation of a mental connection having a relative strength that answers to the relative constancy of the physical connection represented. In conformity with the general law that motion pursues the line of least resistance, and that, other things equal, a line once taken by motion is made a line that will be more readily pursued by future motion; we have seen that the ease with which nervous impressions follow one another, is, other things equal, great in proportion to the number of times they have been repeated together in experience. Hence, corresponding to such an invariable relation as that between the resistance of an object and some extension possessed by it, there arises an indissoluble connection in consciousness; and this connection, being as absolute internally as the answering one is externally, undergoes no further change—the inner relation is in perfect equilibrium with the outer relation. Conversely, it hence happens that to such uncertain relations of phenomena as that between clouds and rain, there arise relations of ideas of a like uncertainty; and if, under given aspects of the sky, the tendencies to infer fair or foul weather, correspond to the frequencies with which fair or foul weather follow such aspects, the accumulation of experiences has balanced the mental sequences and the physical sequences. When it is remembered that between these extremes there are countless orders of external connections having different degrees of constancy, and that during the evolution of intelligence there arise answering internal associations having different degrees of cohesion; it will be seen that there is a progress towards equilibrium between the relations of thought and the relations of things. This equilibration can end only when each relation of things has generated in us a relation of thought, such that on the occurrence of the conditions, the relation in thought arises as certainly as the relation in things. Supposing this state to be reached (which however it can be only in infinite time) experience will cease to produce any further mental evolution—there will have been reached a perfect correspondence between ideas and facts; and the intellectual adaptation of man to his circumstances will be complete. The like general truths are exhibited in the process moral of adaptation; which is a continual approach to equilibrium between the emotions and the kinds of conduct necessitated by surrounding conditions. The connections of feelings and actions, are determined in the same way as the connections of ideas: just as repeating the association of two ideas, facilitates the excitement of the one by the other; so does each discharge of feeling into action, render the subsequent discharge of such feeling into such action more easy. Hence it happens that if an individual is placed permanently in conditions which demand more action of a special kind than has before been requisite, or than is natural to him—if the pressure of the painful feelings which these conditions entail when disregarded, impels him to perform this action to a greater extent—if by every more frequent or more lengthened performance of it under such pressure, the resistance is somewhat diminished; then, clearly, there is an advance towards a balance between the demand for this kind of action and the supply of it. Either in himself, or in his descendants continuing to live under these conditions, enforced repetition must eventually bring about a state in which this mode of directing the energies will be no more repugnant than the various other modes previously natural to the race. Hence the limit towards which emotional modification perpetually tends, and to which it must approach indefinitely near (though it can absolutely reach it only in infinite time) is a combination of desires that correspond to all the different orders of activity which the circumstances of life call for—desires severally proportionate in strength to the needs for these orders of activity; and severally satisfied by these orders of activity. In what we distinguish as acquired habits, and in the moral differences of races and nations produced by habits that are maintained through successive generations, we have countless illustrations of this progressive adaptation; which can cease only with the establishment of a complete equilibrium between constitution and conditions.

Possibly some will fail to see how the equilibrations described in this section, can be classed with those preceding them; and will be inclined to say that what are here set down as facts, are but analogies. Nevertheless such equilibrations are as truly physical as the rest. To show this fully, would require a more detailed analysis than can now be entered on. For the present it must suffice to point out, as before (§ 82), that what we know subjectively as states of consciousness, are, objectively, modes of force; that so much feeling is the correlate of so much motion; that the performance of any bodily action is the transformation of a certain amount of feeling into its equivalent amount of motion; that this bodily action is met by forces which it is expended in overcoming; and that the necessity for the frequent repetition of this action, implies the frequent recurrence of forces to be so overcome. Hence the existence in any individual of an emotional stimulus that is in equilibrium with certain external requirements, is literally the habitual production of a certain specialized portion of nervous energy, equivalent in amount to a certain order of external resistances that are habitually met. And thus the ultimate state, forming the limit towards which Evolution carries us, is one in which the kinds and quantities of mental energy daily generated and transformed into motions, are equivalent to, or in equilibrium with, the various orders and degrees of surrounding forces which antagonize such motions.

§ 135. Each society taken as a whole, displays the process of equilibration in the continuous adjustment of its population to its means of subsistence. A tribe of men living on wild animals and fruits, is manifestly, like every tribe of inferior creatures, always oscillating about that average number which the locality can support. Though by artificial production, and by successive improvements in artificial production, a superior race continually alters the limit which external conditions put to population; yet there is ever a checking of population at the temporary limit reached. It is true that where the limit is being so rapidly changed as among ourselves, there is no actual stoppage: there is only a rhythmical variation in the rate of increase. But in noting the causes of this rhythmical variation—in watching how, during periods of abundance, the proportion of marriages increases, and how it decreases during periods of scarcity; it will be seen that the expansive force produces unusual advance whenever the repressive force diminishes, and *vice versâ*; and thus there is as near a balancing of the two as the changing conditions permit.

The internal actions constituting social functions, exemplify the general principle no less clearly. Supply and demand are continually being adjusted throughout all industrial processes; and this equilibration is interpretable in the same way as preceding ones. The production and distribution of a commodity, is the expression of a certain aggregate of forces causing special kinds and amounts of motion. The price of this commodity, is the measure of a certain other aggregate of forces expended by the labourer who purchases it, in other kinds and amounts of motion. And the variations of price represent a rhythmical balancing of these forces. Every rise or fall in the rate of interest, or change in the value of a particular security, implies a conflict of forces in which some, becoming temporarily predominant, cause a movement that is presently arrested or equilibrated by the increase of opposing forces; and amid these daily and hourly oscillations, lies a more slowly-varying medium, into which the value ever tends to settle; and would settle but for the constant addition of new influences. As in the individual organism so in the social organism, functional equilibrations generate structural equilibrations. When on the workers in any trade there comes an increased demand, and when in return for the increased supply, there is given to them an amount of other commodities larger than was before habitual—when, consequently, the resistances overcome by them in sustaining life are less than the resistances overcome by other workers; there results a flow of other workers into this trade. This flow continues until the extra demand is met, and the wages so far fall again, that the total resistance overcome in obtaining a given amount of produce, is as great in this newly-adopted occupation as in the occupations whence it drew recruits. The occurrence of motion along lines of least resistance, was before shown to necessitate the growth of population in those places where the labour required for self-maintenance is the smallest; and here we further see that those engaged in any such advantageous locality, or advantageous business, must multiply till there arises an approximate balance between this locality or business and others accessible to the same citizens. In determining the career of every youth, we see an estimation by parents of the respective advantages offered by all that are available, and a choice of the one which promises best; and through the consequent influx into trades that are at the time most profitable, and the withholding of recruits from over-stocked trades, there is insured a general equipoise between the power of each social organ and the function it has to perform.

The various industrial actions and re-actions thus continually alternating, constitute a dependent moving equilibrium like that which is maintained among the functions of an individual organism. And this dependent moving equilibrium parallels those already contemplated, in its tendency to become more complete. During early stages of social evolution, while yet the resources of the locality inhabited are unexplored, and the arts of production undeveloped, there is never anything more than a temporary and partial balancing of such actions, under the form of acceleration or retardation of growth. But when a society approaches the maturity of that type on which it is organized, the various industrial activities settle down into a comparatively constant state. Moreover, it is observable that advance in organization, as well as advance in growth, is conducive to a better equilibrium of industrial functions. While the diffusion of mercantile information is slow, and the means of transport deficient, the adjustment of supply to demand is extremely imperfect: great over-production of each commodity followed by great under-production, constitute a rhythm having extremes that depart very widely from the mean state in which demand and supply are equilibrated. But when good roads are made, and there is a rapid diffusion of printed or written intelligence, and still more when railways and telegraphs come into existence—when the periodical fairs of early days lapse into weekly markets, and these into daily markets; there is gradually produced a better balance of production and consumption. Extra demand is much more quickly followed by augmented supply; and the rapid oscillations of price within narrow limits on either side of a comparatively uniform mean, indicate a near approach to equilibrium. Evidently this industrial progress has for its limit, that which Mr. Mill has called "the stationary state." When population shall have become dense over all habitable parts of the globe; when the resources of every region have been fully explored; and when the productive arts admit of no further improvements; there must result an almost complete balance, both between the fertility and mortality of each society, and between its producing and consuming activities. Each society will exhibit only minor deviations from its average number, and the rhythm of its industrial functions will go on from day to day and year to year with comparatively insignificant perturbations. This limit, however, though we are inevitably advancing towards it, is indefinitely remote; and can never indeed be absolutely reached. The peopling of the Earth up to the point supposed, cannot take place by simple spreading. In the future, as in the past, the process will be carried on rhythmically, by waves of emigration from new and higher centres of civilization successively arising; and by the supplanting of inferior races by the superior races they beget; and the process so carried on must be extremely slow. Nor does it seem to me that such an equilibration will, as Mr. Mill suggests, leave scope for further mental culture and moral progress; but rather that the approximation to it must be simultaneous with the approximation to complete equilibrium between man's nature and the conditions of his existence.

One other kind of social equilibration has still to be considered:—that which results in the establishment of governmental institutions, and which becomes complete as these institutions fall into harmony with the desires of the people. There is a demand and supply in political affairs as in industrial affairs; and in the one case as in the other, the antagonist forces produce a rhythm which, at first extreme in its oscillations, slowly settles down into a moving equilibrium of comparative regularity. Those aggressive impulses inherited from the pre-social state—those tendencies to seek self-satisfaction regardless of injury to other beings, which are essential to a predatory life, constitute an anti-social force, tending ever to cause conflict and eventual separation of citizens. Contrariwise, those desires whose ends can be achieved only by union, as well as those sentiments which find satisfaction through intercourse with fellow-men, and those resulting in what we call loyalty, are forces tending to keep the units of a society together. On the one hand, there is in each citizen, more or less of resistance against all restraints imposed on his actions by other citizens: a resistance which, tending continually to widen each individual's sphere of action, and reciprocally to limit the spheres of action of other individuals, constitutes a repulsive force mutually exercised by the members of a social aggregate. On the other hand, the general sympathy of man for man, and the more special sympathy of each variety of man for others of the same variety, together with sundry allied feelings which the social state gratifies, act as an attractive force, tending ever to keep united those who have a common ancestry. And since the resistances to be overcome in satisfying the totality of their desires when living separately, are greater than the resistances to be overcome in satisfying the totality of their desires when living together, there is a residuary force that prevents their separation. Like all other opposing forces, those exerted by citizens on each other, are ever producing alternating movements, which, at first extreme, undergo a gradual diminution on the way to ultimate equilibrium. In small, undeveloped societies, marked rhythms result from these conflicting tendencies. A tribe whose members have held together for a generation or two, reaches a size at which it will not hold together; and on the occurrence of some event causing unusual antagonism among its members, divides. Each primitive nation, depending largely for its continued union on the character of its chief, exhibits wide oscillations between an extreme in which the subjects are under rigid restraint, and an extreme in which the restraint is not enough to prevent disorder. In more advanced nations of like type, we always find violent actions and reactions of the same essential nature—"despotism tempered by assassination," characterizing a political state in which unbearable repression from time to time brings about a bursting of all bonds. In this familiar fact, that a period of tyranny is followed by a period of license and *vice versâ*, we see how these opposing forces are ever equilibrating each other; and we also see, in the tendency of such movements and counter-movements to become more moderate, how the equilibration progresses towards completeness. The conflicts between Conservatism (which stands for the restraints of society over the individual) and Reform (which stands for the liberty of the individual against society), fall within slowly approximating limits; so that the temporary predominance of either, produces a less marked deviation from the medium state. This process, now so far advanced among ourselves that the oscillations are comparatively unobtrusive, must go on till the balance between the antagonist forces approaches indefinitely near perfection. For, as we have already seen, the adaptation of man's nature to the conditions of his existence, cannot cease until the internal forces which we know as feelings are in equilibrium with the external forces they encounter. And the establishment of this equilibrium, is the arrival at a state of human nature and social organization, such that the individual has no desires but those which may be satisfied without exceeding his proper sphere of action, while society maintains no restraints but those which the individual voluntarily respects. The progressive extension of the liberty of citizens, and the reciprocal removal of political restrictions, are the steps by which we advance towards this state. And the ultimate abolition of all limits to the freedom of each, save those imposed by the like freedom of all, must result from the complete equilibration between man's desires and the conduct necessitated by surrounding conditions.

Of course in this case, as in the preceding ones, there is thus involved a limit to the increase of heterogeneity. A few pages back, we reached the conclusion that each advance in mental evolution, is the establishment of some further internal action, corresponding to some further external action—some additional connection of ideas or feelings, answering to some before unknown or unantagonized connection of phenomena. We inferred that each such new function, involving some new modification of structure, implies an increase of heterogeneity; and that thus, increase of heterogeneity must go on, while there remain any outer relations affecting the organism which are unbalanced by inner relations. Whence we saw it to follow that increase of heterogeneity can come to an end only as equilibration is completed. Evidently the like must simultaneously take place with society. Each increment of heterogeneity in the individual, must directly or indirectly involve, as cause or consequence, some increment of heterogeneity in the arrangements of the aggregate of individuals. And the limit to social complexity can be arrived at, only with the establishment of the equilibrium, just described, between social and individual forces.

§ 136. Here presents itself a final question, which has probably been taking a more or less distinct shape in the minds of many, while reading this chapter. "If Evolution of every kind, is an increase in complexity of structure and function that is incidental to the universal process of equilibration—if equilibration, passing through the gradually-perfected forms of moving equilibrium, must end in complete rest; what is the fate towards which all things tend? If the bodies constituting our Solar System are slowly dissipating the forces they possess—if the Sun is losing his heat at a rate which, though insignificant as stated in terms of our chronology, will tell in millions of years—if geologic and meteorologic processes cannot but diminish in activity as the Sun's radiations diminish—if with the diminution of these radiations there must also go on a diminution in the quantity of vegetal and animal existence—if Man and Society, however high the degree of evolution at which they arrive, are similarly dependent on this supply of force that is gradually coming to an end—if thus the highest, equally with the lowest, terrestrial life, must eventually dwindle and disappear; are we not manifestly progressing towards omnipresent death? And have we thus to contemplate, as the out-come of things, a universe of extinct suns round which circle planets devoid of life?"

That such a state must be the proximate end of the processes everywhere going on, seems beyond doubt. But the further question tacitly involved, whether this state will continue eternally, is quite a different one. To give a positive answer to this further question would be quite illegitimate; since to affirm any proposition into which unlimited time enters as one of the terms, is to affirm a proposition of which one term cannot be represented in consciousness—is to affirm an unthinkable proposition. At a first glance it may appear that the reverse conclusion must be equally illegitimate; and that so the question is altogether insoluble. But further consideration will show that this is not true. So long as the terms to which we confine our reasonings are finite, the finite conclusions reached are not necessarily illegitimate. Though, if the general argument, when carried out, left no apparent escape from the inference that the state of rest to which Evolution is carrying things, must, when arrived at, last for ever, this inference would be invalid, as transcending the scope of human intelligence; yet if, on pushing further the general argument, we bring out the inference that such a state will not last for ever, this inference is not necessarily invalid: since, by the hypothesis, it contains no terms necessarily transcending the scope of human intelligence. It is permissible therefore, to inquire, what are the probable ulterior results of this process which must bring Evolution to a close in Universal Death. Without being so rash as to form anything like a positive conclusion on a matter so vast and so far beyond the boundaries of exact science; we may still inquire what *seems* to be the remote future towards which the facts point.

It has been already shown that all equilibration, so far as we can trace it, is relative. The dissipation of a body's motion by communication of it to surrounding matter, solid, liquid, gaseous, and ethereal, tends to bring the body to a fixed position in relation to the matter that abstracts its motion. But all its other motions continue as before. The arrest of a cannon-shot does not diminish its movement towards the East at a thousand miles an hour, along with the wall it has struck; and a gradual dispersion of the Earth's rotatory motion, would abstract nothing from the million and a half miles per day through which the Earth speeds in its orbit. Further, we have to bear in mind that this motion, the disappearance of which causes relative equilibration, is not lost but simply transferred; and by continual division and subdivision finally reduced to ethereal undulations and radiated through space. Whether the sensible motion dissipated during relative equilibration, is directly transformed into insensible motion, as happens in the case of the Sun; or whether, as in the sensible motions going on around us, it is directly transformed into smaller sensible motions, and these into still smaller, until they become insensible, matters not. In every instance the ultimate result is, that whatever motion of masses is lost, re-appears as molecular motion pervading space. Thus the questions we have to consider, are—Whether after the completion of all the relative equilibrations above contemplated as bringing Evolution to a close, there remain any further equilibrations to be effected?—Whether there are any other motions of masses that must eventually be transformed into molecular motion?—And if there are such other motions, what must be the consequence when the molecular motion generated by their transformation, is added to that which already exists?

To the first of these questions the answer is, that there *do* remain motions which are undiminished by all the relative equilibrations thus far considered; namely, the motions of translation possessed by those vast masses of incandescent matter called stars—masses now known to be suns that are in all probability, like our own, surrounded by circling groups of planets. The belief that the stars are literally fixed, has long since been exploded: observation has proved many of them to have sensible proper motions. Moreover, it has been ascertained by measurement, that in relation to the stars nearest to us, our own star is moving at the rate of about half a million miles per day; and if, as is admitted to be not improbable by sundry astronomers, our own star is traversing space in the same direction with adjacent stars, its absolute velocity may be, and most likely is, immensely greater than this. Now no such changes as those taking place within the Solar System, even when carried to the extent of integrating the whole of its matter into one mass, and diffusing all its relative movements in an insensible form through space, can affect these sidereal movements. Hence, there appears no alternative but to infer, that these sidereal movements must remain to be equilibrated by some subsequent process.

The next question that arises, if we venture to inquire the probable nature of this process, is—To what law do sidereal motions conform? And to this question Astronomy replies—the law of gravitation. The relative motions of binary stars have proved this. When it was discovered that certain of the double stars are not optically double but physically double, and move round each other, it was at once suspected that their revolutions might be regulated by a mutual attraction like that which regulates the revolutions of planets and satellites. The requisite measurements having been from time to time made, the periodic times of sundry binary stars were calculated on this assumption; and the subsequent performances of their revolutions in the predicted periods, have completely verified the assumption. If, then, it is demonstrated that these remote bodies are centres of gravitation— if we infer that all other stars are centres of gravitation, as we may fairly do—and if we draw the unavoidable corollary, that this gravitative force which so conspicuously affects stars that are comparatively near each other, must affect remote stars; we find ourselves led to the conclusion that all the members of our Sidereal System gravitate, individually and as an aggregate.

But if these widely-dispersed moving masses mutually gravitate, what must happen? There appears but one tenable answer. Even supposing they were all absolutely equal in weight, and arranged into an annulus with absolute regularity, and endowed with exactly the amounts of centrifugal force required to prevent nearer approach to their common centre of gravity; the condition would still be one which the slightest disturbing force would destroy. Much more then are we driven to the inference, that our actual Sidereal System cannot preserve its present arrangement: the irregularities of its distribution being such as to render even a temporary moving equilibrium impossible. If the stars are so many centres of an attractive force that varies inversely as the square of the distance, there appears to be no escape from the conclusion, that the structure of our galaxy must be undergoing change; and must continue to undergo change.

Thus, in the absence of tenable alternatives, we are brought to the positions:—1, that the stars are in motion;—2, that they move in conformity with the law of gravitation;—3, that, distributed as they are, they cannot move in conformity with the law of gravitation, without undergoing change of arrangement. If now we permit ourselves to take a further step, and ask the nature of this change of arrangement, we find ourselves obliged to infer a progressive concentration. Whether we do or do not suppose the clustering which is now visible, to have been caused by mutual gravitation acting throughout past eras, as the hypothesis of Evolution implies, we are equally compelled to conclude that this clustering must increase throughout future eras. Stars at present dispersed, must become locally aggregated; existing aggregations, at the same time that they are enlarged by the drawing in of adjacent stars, must grow more dense; and aggregations must coalesce with each other: each greater degree of concentration augmenting the force by which further concentration is produced.

And now what must be the limit of this concentration? The mutual attraction of two individual stars, when it so far predominates over other attractions as to cause approximation, almost certainly ends in the formation of a binary star; since the motions generated by other attractions, prevent the two stars from moving in straight lines to their common centre of gravity. Between small clusters, too, having also certain proper motions as clusters, mutual attraction may lead, not to complete union, but to the formation of binary clusters. As the process continues however, and the clusters become larger, it seems clear that they must move more directly towards each other, thus forming clusters of increasing density; and that eventually all clusters must unite into one comparatively close aggregation. While, therefore, during the earlier stages of concentration, the probabilities are immense against the actual contact of these mutually-gravitating masses; it is tolerably manifest, that as the concentration increases, collision must become probable, and ultimately certain. This is an inference not lacking the support of high authority. Sir John Herschel, treating of those numerous and variously-aggregated clusters of stars revealed by the telescope, and citing with apparent approval his father's opinion, that the more diffused and irregular of these, are "globular clusters in a less advanced state of condensation;" subsequently remarks, that "among a crowd of solid bodies of whatever size, animated by independent and partially opposing impulses, motions opposite to each other *must* produce collision, destruction of velocity, and subsidence or near approach towards the centre of preponderant attraction; while those which conspire, or which remain outstanding after such conflicts, *must* ultimately give rise to circulation of a permanent character." Now what is here alleged of these minor sidereal aggregations, cannot be denied of the large aggregations; and thus the above-described process of concentration, appears certain to bring about an increasingly-frequent integration of masses.

We have next to consider the consequences of the accompanying loss of velocity. The sensible motion which disappears, cannot be destroyed; but must be transformed into insensible motion. What will be the effect of this insensible motion? Some approach to a conception of it, will be made by considering what would happen were the comparatively insignificant motion of our planet thus transformed. In his essay on "The Inter-action of Natural Forces," Prof. Helmholtz states the thermal equivalent of the Earth's movement through space; as calculated on the now received datum of Mr. Joule. "If our Earth," he says, "were by a sudden shock brought to rest in her orbit,—which is not to be feared in the existing arrangement of our system—by such a shock a quantity of heat would be generated equal to that produced by the combustion of fourteen such Earths of solid coal. Making the most unfavourable assumption as to its capacity for heat, that is, placing it equal to that of water, the mass of the Earth would thereby be heated 11,200 degrees; it would therefore be quite fused, and for the most part reduced to vapour. If then the Earth, after having been thus brought to rest, should fall into the Sun, which of course would be the case, the quantity of heat developed by the shock would be 400 times greater." Now so relatively small a momentum as that acquired by the Earth in falling through 95,000,000 of miles to the Sun, being equivalent to a molecular motion such as would reduce the Earth to gases of extreme rarity; what must be the molecular motion generated by the mutually-arrested momenta of two stars, that have moved to their common centre of gravity through spaces immeasurably greater? There seems no alternative but to conclude, that this molecular motion must be so great, as to reduce the matter of the stars to an almost inconceivable tenuity—a tenuity like that which we ascribe to nebular matter. Such being the immediate effect of the integration of any two stars in a concentrating aggregate, what must be the ulterior effect on the aggregate as a whole? Sir John Herschel, in the passage above quoted, describing the collisions that must arise in a mutually-gravitating group of stars, adds that those stars "which remain outstanding after such conflicts, *must* ultimately give rise to circulation of a permanent character." The problem, however, is here dealt with purely as a mechanical one: the assumption being, that the mutually-arrested masses will continue as masses—an assumption to which no objection was apparent at the time when Sir John Herschel wrote this passage; since the doctrine of the correlation of forces was not then recognized. But obliged as we now are to conclude, that stars moving at the high velocities acquired during concentration, will, by mutual arrest, be dissipated into gases of great tenuity, the problem becomes different; and a different inference appears unavoidable. For the diffused matter produced by such conflicts, must form a resisting medium, occupying that central region of the aggregate through which its members from time to time pass in describing their orbits—a resisting medium which they cannot move through without having their velocities diminished. Every further such collision, by augmenting this resisting medium, and making the losses of velocity greater, must further aid in preventing the establishment of that equilibrium which would else arise; and so must conspire to produce more frequent collisions. And the nebulous matter thus formed, presently enveloping and extending beyond the whole aggregate, must, by continuing to shorten their gyrations, entail an increasingly-active integration and re-active disintegration of the moving masses; until they are all finally dissipated. This, indeed, is the conclusion which, leaving out all consideration of the process gone through, presents itself as a simple deduction from the persistence of force. If the stars have been, and still are, concentrating however indirectly on their common centre of gravity, and must eventually reach it; it is a corollary from the persistence of force, that the quantities of motion they have severally acquired, must suffice to carry them away from the common centre of gravity to those remote regions whence they originally began to move towards it. And since, by the conditions of the case, they cannot return to these remote regions in the shape of concrete masses, they must return in the shape of diffused masses. Action and reaction being equal and opposite, the momentum producing dispersion, must be as great as the momentum acquired by aggregation; and being spread over the same quantity of matter, must cause an equivalent distribution through space, whatever be the form of the matter. One condition, however, essential to the literal fulfilment of this result, must be specified; namely, that the quantity of molecular motion produced and radiated into space by each star in the course of its formation from diffused matter, shall be compensated by an equal quantity of molecular motion radiated from other parts of space into the space which our Sidereal System occupies. In other words, if we set out with that amount of molecular motion implied by the existence of the matter of our Sidereal System in a nebulous form; then it follows from the persistence of force, that if this matter undergoes the re-distribution constituting Evolution, the quantity of molecular motion given out during the integration of each mass, plus the quantity of molecular motion given out during the integration of all the masses, must suffice again to reduce it to the same nebulous form. Here indeed we arrive at an impassable limit to our reasonings; since we cannot know whether this condition is or is not fulfilled. On the hypothesis of an unlimited space, containing, at certain intervals, Sidereal Systems like our own, it may be that the quantity of molecular motion radiated into the region occupied by our Sidereal System, is equal to that which our Sidereal System radiates; in which case the quantity of motion possessed by it, remaining undiminished, our Sidereal System may continue during unlimited time, to repeat this alternate concentration and diffusion. But if, on the other hand, throughout boundless space there exist no other Sidereal Systems subject to like changes, or if such other Sidereal Systems exist at more than a certain average distance from each other; then it seems an unavoidable conclusion that the quantity of motion possessed, must diminish by radiation into unoccupied space; and that so, on each successive resumption of the nebulous form, the matter of our Sidereal System will occupy a less space; until at the end of an infinite time it reaches either a state in which its concentrations and diffusions are relatively small, or a state of complete aggregation and rest. Since, however, we have no evidence showing the existence or non-existence of Sidereal Systems throughout remote space; and since, even had we such evidence, a legitimate conclusion could not be drawn from premises of which one element (unlimited space) is inconceivable; we must be for ever without answer to this transcendent question. All we can say is, that so far as the data enable us to judge, the integration of our Sidereal System will be followed by disintegration; that such integration and disintegration will be repeated; and that, for anything we know to the contrary, the alternation of them may continue without limit.

But leaving this ultimate insoluble problem, and confining ourselves to the proximate and not necessarily insoluble one, we find reason for thinking that after the completion of those various equilibrations which bring to a close all the forms of Evolution we have contemplated, there must still continue an equilibration of a far wider kind. When that integration everywhere in progress throughout our Solar System, has reached its climax, there will remain to be effected the immeasurably greater integration of our Solar System, with all other such systems. As in those minor forms now going on around us, this integration with its concomitant equilibration, involves the change of aggregate motion into diffused motion; so in those vaster forms hereafter to be carried out, there must similarly be gained in molecular motion what is lost in the motion of masses; and the inevitable transformation of this motion of masses into molecular motion, cannot take place without reducing the masses to a nebulous form. Thus we seem led to the conclusion that the entire process of things, as displayed in the aggregate of the visible Universe, is analogous to the entire process of things as displayed in the smallest aggregates. Where, as in organic bodies, the whole series of changes constituting Evolution can be traced, we saw that, dynamically considered, Evolution is a change from molecular motion to the motion of masses; and this change, becoming more active during the ascending phase of Evolution while the masses increase in bulk and heterogeneity, eventually begins to get less active; until, passing through stages in which the integration grows greater, and the equilibrium more definite, it finally ceases; whereupon there arises, by an ulterior process, an increase of molecular motion, ending in the more or less complete dissolution of the aggregate. And here we find reason to believe that, along with each of the thousands of similar ones dispersed through the heavens, our Solar System, after passing through stages during which the motion of masses is produced at the expense of lost molecular motion, and during which there goes on an increasingly active differentiation and integration, arrives at a climax whence these changes, beginning to decline in activity, slowly bring about that complete integration and equilibration which in other cases we call death; and that there afterwards comes a time, when the still-remaining motions of masses are transformed into a molecular motion which causes dissolution of the masses. Motion as well as Matter being fixed in quantity, it would seem that the change in the distribution of Matter which Motion effects, coming to a limit in whichever direction it is carried, the indestructible Motion thereupon necessitates a reverse distribution.

Apparently, the universally-coexistent forces of attraction and repulsion, which, as we have seen, necessitate rhythm in all minor changes throughout the Universe, also necessitate rhythm in the totality of its changes—produce now an immeasurable period during which the attractive forces predominating, cause universal concentration, and then an immeasurable period during which the repulsive forces predominating, cause universal diffusion—alternate eras of Evolution and Dissolution. And thus there is suggested the conception of a past during which there have been successive Evolutions similar to that which is now going on; and a future during which successive other such Evolutions may go on.

Let none suppose, however, that this is to be taken as anything more than a speculation. In dealing with times and spaces and forces so immensely transcending those of which we have definite experience, we are in danger of passing the limits to human intelligence. Though these times and spaces and forces cannot literally be classed as infinite; yet they are so utterly beyond the possibility of definite conception, as to be almost equally unthinkable with the infinite. What has been above said, should therefore be regarded simply as a possible answer to a possible doubt. When, pushing to its extreme the argument that Evolution must come to a close in complete equilibrium or rest, the reader suggests that for aught which appears to the contrary, the Universal Death thus implied will continue indefinitely; it is legitimate to point out how, on carrying the argument still further, we are led to infer a subsequent Universal Life. But while this last inference may fitly be accepted as a demurrer to the first, it would be unwise to accept it in any more positive sense.

§ 137. Returning from this parenthetical discussion, concerning the probable or possible state of things that may arise after Evolution has run its course; and confining ourselves to the changes constituting Evolution, with which alone we are immediately concerned; we have now to inquire whether the cessation of these changes, in common with all their transitional characteristics, admits of *à priori* proof. It will soon become apparent that equilibration, not less than the preceding general principles, is deducible from the persistence of force.

We have seen (§ 85) that phenomena are interpretable only as the results of universally-coexistent forces of attraction and repulsion. These universally-coexistent forces of attraction and repulsion, are, indeed, the complementary aspects of that absolutely persistent force which is the ultimate datum of consciousness. Just in the same way that the equality of action and re-action is a corollary from the persistence of force, since their inequality would imply the disappearance of the differential force into nothing, or its appearance out of nothing; so, we cannot become conscious of an attractive force without becoming simultaneously conscious of an equal and opposite repulsive force. For every experience of a muscular tension, (under which form alone we can immediately know an attractive force,) presupposes an equivalent resistance—a resistance shown in the counter-balancing pressure of the body against neighbouring objects, or in that absorption of force which gives motion to the body, or in both—a resistance which we cannot conceive as other than equal to the tension, without conceiving force to have either appeared or disappeared, and so denying the persistence of force. And from this necessary correlation, results our inability, before pointed out, of interpreting any phenomena save in terms of these correlatives—an inability shown alike in the compulsion we are under to think of the statical forces which tangible matter displays, as due to the attraction and repulsion of its atoms, and in the compulsion we are under to think of dynamical forces exercised through space, by regarding space as filled with atoms similarly endowed. Thus from the existence of a force that is for ever unchangeable in quantity, there follows, as a necessary corollary, the co-extensive existence of these opposite forms of force—forms under which the conditions of our consciousness oblige us to represent that absolute force which transcends our knowledge.

But the forces of attraction and repulsion being universally co-existent, it follows, as before shown, that all motion is motion under resistance. Units of matter, solid, liquid, aëriform, or ethereal, filling the space which any moving body traverses, offer to such body the resistance consequent on their cohesion, or their inertia, or both. In other words, the denser or rarer medium which occupies the places from moment to moment passed through by such moving body, having to be expelled from them, as much motion is abstracted from the moving body as is given to the medium in expelling it from these places. This being the condition under which all motion occurs, two corollaries result. The first is, that the deductions perpetually made by the communication of motion to the resisting medium, cannot but bring the motion of the body to an end in a longer or shorter time. The second is, that the motion of the body cannot cease until these deductions destroy it. In other words, movement must continue till equilibration takes place; and equilibration must eventually take place. Both these are manifest deductions from the persistence of force. To say that the whole or part of a body's motion can disappear, save by transfer to something which resists its motion, is to say that the whole or part of its motion can disappear without effect; which is to deny the persistence of force. Conversely, to say that the medium traversed can be moved out of the body's path, without deducting from the body's motion, is to say that motion of the medium can arise out of nothing; which is to deny the persistence of force. Hence this primordial truth is our immediate warrant for the conclusions, that the changes which Evolution presents, cannot end until equilibrium is reached; and that equilibrium must at last be reached.

Equally necessary, because equally deducible from this same truth that transcends proof, are the foregoing propositions respecting the establishment and maintenance of moving equilibria, under their several aspects. It follows from the persistence of force, that the various motions possessed by any aggregate, either as a whole or among its parts, must be severally dissipated by the resistances they severally encounter; and that thus, such of them as are least in amount, or meet with greatest opposition, or both, will be brought to a close while the others continue. Hence in every diversely moving aggregate, there results a comparatively early dissipation of motions which are smaller and much resisted; followed by long-continuance of the larger and less-resisted motions; and so there arise dependent and independent moving equilibria. Hence also may be inferred the tendency to conservation of such moving equilibria; since, whenever the new motion given to the parts of a moving equilibrium by a disturbing force, is not of such kind and amount that it cannot be dissipated before the pre-existing motions (in which case it brings the moving equilibrium to an end) it must be of such kind and amount that it can be dissipated before the pre-existing motions (in which case the moving equilibrium is re-established).

Thus from the persistence of force follow, not only the various direct and indirect equilibrations going on around, together with that cosmical equilibration which brings Evolution under all its forms to a close; but also those less manifest equilibrations shown in the re-adjustments of moving equilibria that have been disturbed. By this ultimate principle is proveable the tendency of every organism, disordered by some unusual influence, to return to a balanced state. To it also may be traced the capacity, possessed in a slight degree by individuals, and in a greater degree by species, of becoming adapted to new circumstances. And not less does it afford a basis for the inference, that there is a gradual advance towards harmony between man's mental nature and the conditions of his existence. After finding that from it are deducible the various characteristics of Evolution, we finally draw from it a warrant for the belief, that Evolution can end only in the establishment of the greatest perfection and the most complete happiness.

18.   Sir David Brewster has recently been citing with approval, a calculation by M. Babinet, to the effect that on

the hypothesis of nebular genesis, the matter of the Sun, when it filled the Earth's orbit, must have taken 3181 years to rotate; and that therefore the hypothesis cannot be true. This calculation of M. Babinet may pair-off with that of M. Comte, who, contrariwise, made the time of this rotation agree very nearly with the Earth's period of revolution round the Sun; for if M. Comte's calculation involved a *petitio principii*, that of M. Babinet is manifestly based on two assumptions, both of which are gratuitous, and one of them totally inconsistent with the doctrine to be tested. He has evidently proceeded on the current supposition respecting the Sun's internal density, which is not proved, and from which there are reasons for dissenting; and he has evidently taken for granted that all parts of the nebulous spheroid, when it filled the Earth's orbit, had the same angular velocity; whereas if (as is implied in the nebular hypothesis, rationally understood) this spheroid resulted from the concentration of far more widely-diffused matter, the angular velocity of its equatorial portion would obviously be immensely greater than that of its central portion.

19. See paper "On the Inter-action of Natural Forces," by Prof. Helmholtz, translated by Prof. Tyndall, and published in the *Philosophical Magazine*, supplement to Vol. XI. fourth series.

20. Until I recently consulted his "Outlines of Astronomy" on another question, I was not aware that so far back as 1833, Sir John Herschel had enunciated the doctrine that "the sun's rays are the ultimate source of almost every motion which takes place on the surface of the earth." He expressly includes all geologic, meteorologic, and vital actions; as also those which we produce by the combustion of coal. The late George Stephenson appears to have been wrongly credited with this last idea.

CHAPTER XVII
SUMMARY AND CONCLUSION

§ 138. In the chapter on "Laws in general," after delineating the progress of mankind in recognizing uniformities of relation among surrounding phenomena—after showing how the actual succession in the establishment of different orders of co-existences and sequences, corresponds with the succession deducible *à priori* from the conditions to human knowledge— after showing how, by the ever-multiplying experiences of constant connections among phenomena, there has been gradually generated the conception of universal conformity to law; it was suggested that this conception will become still clearer, when it is perceived that there are laws of wider generality than any of those at present accepted.

The existence of such more general laws, is, indeed, almost implied by the *ensemble* of the facts set forth in the above-named chapter; since they make it apparent, that the process hitherto carried on, of bringing phenomena under fewer and wider laws, has not ceased, but is advancing with increasing rapidity. Apart, however, from evidence of this kind, the man of science, hourly impressed with new proof of uniformity in the relations of things, until the conception of uniformity has become with him a necessity of thought, tacitly entertains the conclusion that the minor uniformities which Science has thus far established, will eventually be merged in uniformities that are universal. Taught as he is by every observation and experiment, to regard phenomena as manifestations of Force; and learning as he does to contemplate Force as unchangeable in amount; there tends to grow up in him a belief in unchangeable laws common to Force under all its manifestations. Though he may not have formulated it to himself, he is prepared to recognize the truth, that, being fixed in quantity, fixed in its two ultimate modes of presentation (Matter and Motion), and fixed in the conditions under which it is presented (Time and Space); Force must have certain equally fixed laws of action, common to all the changes it produces.

Hence to the classes who alone are likely to read these pages, the hypothesis of a fundamental unity, extending from the simplest inorganic actions up to the most complex associations of thought and the most involved social processes, will have an *à priori* probability. All things being recognized as having one source, will be expected to exhibit one method. Even in the absence of a clue to uniformities co-extensive with all modes of Force, as the mathematical uniformities are co-extensive with Space and Time, it will be inferred that such uniformities exist. And thus a certain presumption will result in favour of any formula, of a generality great enough to include concrete phenomena of every order.

§ 139. In the chapters on the "Law of Evolution," there was set forth a principle, which, so far as accessible evidence enables us to judge, possesses this universality. The order of material changes, first perceived to have certain constant characteristics in cases where it could be readily traced from beginning to end, we found to have these same characteristics in cases where it could be less readily traced; and we saw numerous indications that these same characteristics were displayed during past changes of which we have no direct knowledge. The transformation of the homogeneous into the heterogeneous, first observed by naturalists to be exhibited during the development of every plant and animal, proved to be also exhibited during the development of every society; both in its political and industrial organization, and in all the products of social life,—language, science, art, and literature. From the disclosures of geology, we drew adequate support for the conclusion, that in the structure of the Earth there has similarly been a progress from uniformity, through ever-increasing degrees of multiformity, to the complex state which we now see. And on the assumption of that nebular origin to which so many facts point, we inferred that a like transition from unity to variety of distribution, must have been undergone by our Solar System; as well as by that vast assemblage of such systems constituting the visible Universe. This definition of the metamorphosis, first asserted by physiologists of organic aggregates only, but which we thus found reason to think, holds of all other aggregates, proved on further inquiry to be too wide. Its undue width was shown to arise from the omission of certain other characteristics, that are, not less than the foregoing one, displayed throughout all kinds of Evolution. We saw that simultaneously with the change from homogeneity to heterogeneity, there takes place a change from indefiniteness of arrangement to definiteness of arrangement—a change everywhere equally traceable with that which it accompanies. Further consideration made it apparent, that the increasing definiteness thus manifested along with increasing heterogeneity, necessarily results from increasing integration of the parts severally rendered unlike. And thus we finally reached the conclusion, that there has been going on throughout an immeasurable past, is still going on, and will continue to go on, an advance from a diffused, indeterminate, and uniform distribution of Matter, to a concentrated, determinate, and multiform distribution of it.

At a subsequent stage of our inquiry, we discovered that this progressive change in the arrangement of Matter, is accompanied by a parallel change in the arrangement of Motion—that every increase in the structural complexity of things, involves a corresponding increase in their functional complexity. It was shown that along with the integration of molecules into masses, there arises an integration of molecular motion into the motion of masses; and that as fast as there results variety in the sizes and forms of aggregates and their relations to incident forces, there also results variety in their movements. Whence it became manifest, that the general process of things is from a confused simplicity to an orderly complexity, in the distribution of both Matter and Motion.

It was pointed out, however, that though this species of transformation is universal, in the sense of holding throughout all classes of phenomena, it is not universal in the sense of being continued without limit in all classes of phenomena. Those aggregates which exhibit the entire change from uniformity to multiformity of structure and function, in comparatively short periods, eventually show us a reverse set of changes: Evolution is followed by Dissolution. The differentiations and integrations of Matter and Motion, finally reach a degree which the conditions do not allow them to pass; and there then sets in a process of disintegration and assimilation, of both the parts and the movements that were before growing more united and more distinct.

But under one or the other of these processes, all observable modifications in the arrangement of things may be classed. Every change comes under the head of integration or disintegration, material or dynamical; or under the head of differentiation or assimilation, material or dynamical; or under both. Each inorganic mass is either undergoing increase by the combination with it of surrounding elements for which its parts have affinity; or undergoing decrease by the solvent and abraiding action of surrounding elements; or both one and the other in varied succession and combination. By perpetual additions and losses of heat, it is having its parts temporarily differentiated from each other, or temporarily assimilated to each other, in molecular state. And through the actions of divers agents, it is also undergoing certain permanent molecular re-arrangements; rendering it either more uniform or more multiform in structure. These opposite kinds of change, thus vaguely typified in every surrounding fragment of matter, are displayed in all aggregates with increasing distinctness in proportion as the conditions essential to re-arrangement of parts are fulfilled. So that universally, the process of things is either in the one direction or the other. There is in all cases going on that ever-complicating distribution of Matter and Motion which we call Evolution; save in those cases where it has been brought to a close and reversed by what we call Dissolution.

§ 140. Whether this omnipresent metamorphosis admits of interpretation, was the inquiry on which we next entered. Recognizing the changes thus formulated as consisting in Motions of Matter that are produced by Force, we saw that if they are interpretable at all, it must be by the affiliation of them on certain ultimate laws of Matter, Motion, and Force. We therefore proceeded to inquire what these ultimate laws are.

We first contemplated under its leading aspects, the principle of correlation and equivalence among forces. The genesis of sensible motion by insensible motion, and of insensible motion by sensible motion, as well as the like reciprocal production of those forms of insensible motion which constitute Light, Heat, Electricity, Magnetism, and Chemical Action, was shown to be a now accepted doctrine, that involves certain corollaries respecting the processes everywhere going on around us. Setting out with the probability that the insensible motion radiated by the Sun, is the transformed product of the sensible motion lost during the progressive concentration of the solar mass; we saw that by this insensible motion, are in turn produced the various kinds of sensible motion on the Earth's surface. Besides the inorganic terrestrial changes, we found that the changes constituting organic life are thus originated. We were obliged to conclude that within this category, come the vital phenomena classed as mental, as well as those classed as physical. And it appeared inevitably to follow that of social changes, too, the like must be said. We next saw that phenomena being cognizable by us only as products of Force, manifested under the two-fold form of attraction and repulsion, there results the general law that all Motion must occur in the direction of least resistance, or in the direction of greatest traction, or in the direction of their resultant. It was pointed out that this law is every instant illustrated in the movements of the celestial bodies. The innumerable transpositions of matter, gaseous, liquid, and solid, going on over the Earth's surface, were shown to conform to it. Evidence was given that this same ultimate principle of motion underlies the structural and functional changes of organisms. Throughout the succession of those nervous actions which constitute thought and feeling, as also in the discharge of feeling into action, we no less found this principle conspicuous. Nor did we discover any exception to it in the movements, temporary and permanent, that go on in societies. From the universal coexistence of opposing forces, there also resulted the rhythm of motion. It was shown that this is displayed from the infinitesimal vibrations of molecules up to the enormous revolutions and gyrations of planets; that it is traceable throughout all meteorologic and geologic changes; that the functions of every organic body exemplify it in various forms; that mental activities too, intellectual and emotional, exhibit periodicities of sundry kinds; and that actions and reactions illustrating this law under a still more complex form, pervade social processes.

Such being the principles to which conform all changes produced by Force on the distribution of Matter, and all changes re-actively produced by Matter on the distribution of Force, we proceeded to inquire what must be the consequent nature of any re-distributions produced: having first noted the limiting conditions between which such re-distributions are possible, and the medium conditions that are most favourable to them. The first conclusion arrived at, was, that any finite homogeneous aggregate must inevitably lose its homogeneity, through the unequal exposure of its parts to incident forces. We observed how this was shown in surrounding things, by the habitual establishment of differences between inner and outer parts, and parts otherwise dissimilarly circumstanced. It was pointed out that the production of diversities of structure by forces acting under diverse conditions, has been illustrated in astronomic evolution, supposing such evolution to have taken place; and that a like connection of cause and effect is seen in the large and small modifications undergone by our globe. In the early changes of organic germs, we discovered further evidence that unlikenesses of structure follow unlikenesses of relations to surrounding agencies—evidence enforced by the tendency of the differently-placed members of each species to diverge into varieties. We found that the principle is also conformed to in the establishment of distinctions among our ideas; and that the contrasts, political and industrial, that arise between the parts of societies are no less in harmony with it. The instability of the homogeneous thus caused, and thus everywhere exemplified, we also saw must hold of the unlike parts into which any uniform whole lapses; and that so the less heterogeneous must tend continually to become more heterogeneous—an inference which we also found to be everywhere confirmed by fact. Carrying a step further our inquiry into these actions and reactions between Force and Matter, there was disclosed a secondary cause of increasing multiformity. Every differentiated part becomes, we found, a parent of further differentiations; not only in the sense that it must lose its own homogeneity in heterogeneity, but also in the sense that it must, in growing unlike other parts, become a centre of unlike reactions on incident forces; and by so adding to the diversity of forces at work, must add to the diversity of effects produced. This multiplication of effects, likewise proved to be manifest throughout all Nature. That forces modified in kind and direction by every part of every aggregate, are gradually expended in working changes that grow more numerous and more varied as the forces are subdivided, is shown in the actions and reactions going on throughout the Solar System, in the never-ceasing geologic complications, in the involved symptoms produced in organisms by disturbing influences, in the many thoughts and feelings generated by single impressions, and in the ever-ramifying results of each new agency brought to bear on a society. To which add the corollary, confirmed by abundant facts, that the multiplication of effects must increase in a geometrical progression, as the heterogeneity increases. Completely to interpret the structural changes constituting Evolution, there remained to assign a reason for that increasingly-distinct demarcation of parts, which accompanies the production of differences between parts. This reason we discovered to be, the segregation of mixed units under the action of forces capable of moving them. We saw that when the parts of an aggregate have been made qualitatively unlike by unlike incident forces—that is, when they have become contrasted in the natures of their component units; there necessarily arises a tendency to separation of the dissimilar orders of units from each other, and to aggregation of those units which are similar. This cause of the integration that accompanies differentiation, turned out to be likewise exemplified by all kinds of Evolution—by the formation of celestial bodies, by the moulding of the Earth's crust, by organic modifications, by the establishment of mental distinctions, by the genesis of social divisions. And we inferred, what we may everywhere see, that the segregation thus produced goes on so long as there remains a possibility of making it more complete. At length, to the query whether the processes thus traced out have any limit, there came the answer that they must end in equilibration. That continual division and subdivision of forces, which is instrumental in changing the uniform into the multiform and the multiform into the more multiform, we saw to be at the same time a process by which force is perpetually dissipated; and that dissipation, continuing as long as there remains any force unbalanced by an opposing force, must end in rest. It was shown that when, as happens with aggregates of various orders, a number of movements are going on in combination, the earlier dispersion of the smaller and more resisted movements, entails the establishment of different kinds of moving equilibria: forming transitional stages on the way to complete equilibrium. And further inquiry made it apparent that for the same reason, these moving equilibria have a certain self-conserving power; shown in the neutralization of perturbations, and the adjustment to new conditions. This general principle, like the preceding ones, proved to be traceable throughout all forms of Evolution—astronomic, geologic, biologic, mental and social. And our concluding inference was, that the penultimate stage of this process, in which the extremest degree of multiformity and completest form of moving equilibrium is established, must be one implying the highest conceivable state of humanity.

Thus it became apparent that this transformation of on indefinite, incoherent homogeneity into a definite coherent heterogeneity, which goes on everywhere, until it brings about a reverse transformation, is consequent on certain simple laws of force. Given those universal modes of action which are from moment to moment illustrated in the commonest changes around us, and it follows that there cannot but result the observed metamorphosis of an indeterminate uniformity into a determinate multiformity.

§ 141. Finally, we have asked whether, for these universal modes of action, any common cause is assignable—whether these wide truths are dependent on any single widest truth. And to this question we found a positive answer. These several principles are corollaries from that primordial principle which transcends human intelligence by underlying it.

In the first part of this work it was shown, by analysis of both our religious and our scientific ideas, that while knowledge of the cause which produces effects on our consciousness is impossible, the existence of a cause for these effects is a datum of consciousness. Though Being is cognizable by us only under limits of Time and Space, yet Being without limits of Time and Space was proved to be the indefinite cognition forming the necessary basis of our definite cognitions. We saw that the belief in an Omnipresent Power of which no commencement or cessation can be conceived, is that fundamental element in Religion which survives all its changes of form. We saw that all Philosophies avowedly or tacitly recognize this same ultimate truth:— that while the Relativist rightly repudiates those definite assertions which the Absolutist makes respecting real existence, he is yet at last compelled to unite with him in predicating real existence. And this inexpugnable consciousness in which Religion and Philosophy are at one with Common Sense, proved to be likewise that on which all exact Science is founded. We found that subjective Science can give no account of those conditioned modes of existence which constitute consciousness, without postulating unconditioned existence. And we found that objective Science can give no account of the existence which we know as external, without regarding its changes of form as manifestations of an existence that continues constant under all forms. Absolute Being, or Being which persists without beginning or end, was shown to be the common datum of all human thought; for the sufficient reason that the consciousness of it cannot be suppressed, without the suppression of consciousness itself.

From this truth which transcends proof, we have seen that the general principles above set down, are deducible. That the power or force manifested to us in all phenomena, continues unaltered in quantity, however its mode of manifestation be altered, is a proposition in which these several propositions are involved. It was shown that on the Persistence of Force are based the demonstrations that Matter is indestructible and Motion continuous. When its proofs were examined, the correlation and equivalence of forces was found to follow from the Persistence of Force. The necessity we are under of conceiving Force under the two-fold form of attraction and repulsion, turns out to be but an implication of the necessity we are under of conceiving Force as persistent. On the Persistence of Force, we saw that the law of direction of Motion is dependent; and from it also we saw that the rhythm of Motion necessarily results. Passing to those changes of distribution which, by the Motion it generates, Force produces in Matter, it was pointed out that from the Persistence of Force are severally deducible, the instability of the homogeneous, the multiplication of effects, and that increasing definiteness of structure to which continuous differentiation and integration leads. And lastly we saw that Force being persistent, Evolution cannot cease until equilibrium is reached; and that equilibrium must eventually be reached.

So that given Force manifested in Time and Space, under the forms of Matter and Motion; and it is demonstrable, *à priori*, that there must go on such transformations as we find going on.

§ 142. See then the accumulation of proofs. The advance of human intelligence in establishing laws continually wider in generality, raises the presumption that there are all-comprehensive laws. Turning to the facts, we discern a pervading uniformity in the general course of things where this can be watched, and indications of such uniformity where it cannot be watched. Considering this uniformity analytically, we find it to result from certain simpler uniformities in the actions of Force. And these uniformities prove to be so many necessary implications of that primordial truth which underlies all knowledge—the Persistence of Force. The aspect of things raises a presumption; extended observations lead to an induction that fulfils this presumption; this induction is deductively confirmed; and the laws whence it is deduced are corollaries from that datum without which thought is impossible.

No higher degree of verification than this can be imagined. An induction based on facts so numerous and varied, and falling short of universality only where the facts are beyond observation, possesses of itself a validity greater than that of most scientific inductions. When it is shown that the proposition thus arrived at *à posteriori*, may also be arrived at *à priori*, starting from certain simple laws of force; it is raised to a level with those generalizations of concrete science which are accepted as proved. And when these simple laws of force are affiliated upon that ultimate truth which transcends proof; this dependent proposition takes rank with those propositions of abstract science which are our types of the greatest conceivable certainty.

Let no one suppose that any such degree of certainty is alleged of the various minor propositions brought in illustration of the general argument. Such an assumption would be so manifestly absurd, that it seems scarcely needful to disclaim it. But the truth of the doctrine as a whole, is unaffected by errors in the details of its presentation. As the first principles of mathematics are not invalidated by mistakes made in working out particular equations; so the first principles set forth in the foregoing pages, do not stand or fall with each special statement made in them. If it can be shown that the Persistence of Force is not a datum of consciousness; or if it can be shown that the several laws of force above specified are not corollaries from it; then, indeed, it will be shown that the theory of Evolution has not the certainty here claimed for it. But nothing short of this can invalidate the general conclusions arrived at.

§ 143. If these conclusions be accepted—if it be admitted that they inevitably follow from the truth transcending all others in authority—if it be agreed that the phenomena going on everywhere are parts of the general process of Evolution, save where they are parts of the reverse process of Dissolution; then it must be inferred that all phenomena receive their complete interpretation, only when recognized as parts of these processes. Regarded from the point of view here reached, each change that takes place, is an incident in the course of the ever-complicating distribution of Matter and Motion, except where it is an incident in the course of the reverse distribution; and each such change is fully understood, only when brought under those universal principles of change, to which these transformations necessarily conform. Whence, indeed, it appears to be an unavoidable conclusion, that the limit towards which Science is advancing, must be reached when these formulæ are made all-comprehensive. Manifestly, the perfection of Science, is a state in which all phenomena are seen to be necessary implications of the Persistence of Force. In such a state, the dependence of each phenomenon on the Persistence of Force, must be proved either directly or indirectly—either by showing that it is a corollary of the Persistence of Force, or by showing that it is a corollary from some general proposition deduced from the Persistence of Force. And since all phenomena are incidents in the re-distributions of Matter and Motion; and since there are certain general principles, deducible from the Persistence of Force, to which all these re-distributions conform; it seems inferrable that ultimately all phenomena, where not classed as consequences of the Persistence of Force, must be classed as consequences of these derivative principles.

§ 144. Of course this development of Science into an organized aggregate of direct and indirect deductions from the Persistence of Force, can be achieved only in the remote future; and indeed cannot be completely achieved even then. Scientific progress, is progress in that equilibration of thought and things which we saw is going on, and must continue to go on; but which cannot arrive at perfection in any finite period, because it advances more slowly the further it advances. But though Science can never be entirely reduced to this form; and though only at a far distant time can it be brought nearly to this form; yet much may even now be done in the way of rude approximation. Those who are familiar with the present aspects of Science, must recognize in them the broken outlines of a general organization. The possibility of arranging the facts already accumulated, into the order rudely exhibited in the foregoing pages, will itself incline them to the belief that our knowledge may be put into a more connected shape than it at present has. They will see the probability that many now isolated inductions, may be reduced to the form of deductions from first principles. They will suspect that inferences drawn from the ultimate laws of force, will lead to the investigation and generalization of classes of facts hitherto unexamined. And they will feel, not only that a greater degree of certainty must be acquired by Science, as fast as its propositions are directly or indirectly deduced from the highest of all truths; but also that it must so be rendered a more efficient agent of further inquiry.

To bring scientific knowledge to such degree of logical coherence as is at present possible, is a task to be achieved only by the combined efforts of many. No one man can possess that encyclopedic information required for rightly arranging even the truths already established. But as progress is effected by increments—as all organization, beginning in faint and blurred outlines, is completed by successive modifications and additions; advantage may accrue from an attempt, however rude, to reduce the facts already accumulated—or rather certain classes of them—to something like co-ordination. Such must be the plea for the several volumes which are to succeed this.

§ 145. A few closing words must be said, concerning the general bearings of the doctrines that are now to be further developed. Before proceeding to interpret the detailed phenomena of Life, and Mind, and Society, in terms of Matter, Motion, and Force, the reader must be reminded in what sense the interpretations are to be accepted. In spite of everything said at the outset, there are probably some who have gained the impression that those most general truths set forth in the preceding chapters, together with the truths deducible from them, claim to be something more than relative truths. And, notwithstanding all evidence to the contrary, there will probably have arisen in not a few minds, the conviction that the solutions which have been given, along with those to be derived from them, are essentially materialistic. Let none persist in these misconceptions.

As repeatedly shown in various ways, the deepest truths we can reach, are simply statements of the widest uniformities in our experience of the relations of Matter, Motion, and Force; and Matter, Motion, and Force are but symbols of the Unknown Reality. That Power of which the nature remains for ever inconceivable, and to which no limits in Time or Space can be imagined, works in us certain effects. These effects have certain likenesses of kind, the most general of which we class together under the names of Matter, Motion, and Force; and between these effects there are likenesses of connection, the most constant of which we class as laws of the highest certainty. Analysis reduces these several kinds of effect to one kind of effect; and these several kinds of uniformity to one kind of uniformity. And the highest achievement of Science is the interpretation of all orders of phenomena, as differently-conditioned manifestations of this one kind of effect, under differently-conditioned modes of this one kind of uniformity. But when Science has done this, it has done nothing more than systematize our experience; and has in no degree extended the limits of our experience. We can say no more than before, whether the uniformities are as absolutely necessary, as they have become to our thought relatively necessary. The utmost possibility for us, is an interpretation of the process of things as it presents itself to our limited consciousness; but how this process is related to the actual process, we are unable to conceive, much less to know.

Similarly, it must be remembered that while the connection between the phenomenal order and the ontological order is for ever inscrutable; so is the connection between the conditioned forms of being and the unconditioned form of being, for ever inscrutable. The interpretation of all phenomena in terms of Matter, Motion, and Force, is nothing more than the reduction of our complex symbols of thought, to the simplest symbols; and when the equation has been brought to its lowest terms the symbols remain symbols still. Hence the reasonings contained in the foregoing pages, afford no support to either of the antagonist hypotheses respecting the ultimate nature of things. Their implications are no more materialistic than they are spiritualistic; and no more spiritualistic than they are materialistic. Any argument which is apparently furnished to either hypothesis, is neutralized by as good an argument furnished to the other. The Materialist, seeing it to be a necessary deduction from the law of correlation, that what exists in consciousness under the form of feeling, is transformable into an equivalent of mechanical motion, and by consequence into equivalents of all the other forces which matter exhibits; may consider it therefore demonstrated that the phenomena of consciousness are material phenomena. But the Spiritualist, setting out with the same data, may argue with equal cogency, that if the forces displayed by matter are cognizable only under the shape of those equivalent amounts of consciousness which they produce, it is to be inferred that these forces, when existing out of consciousness, are of the same intrinsic nature as when existing in consciousness; and that so is justified the spiritualistic conception of the external world, as consisting of something essentially identical with what we call mind. Manifestly, the establishment of correlation and equivalence between the forces of the outer and the inner worlds, may be used to assimilate either to the other; according as we set out with one or other term. But he who rightly interprets the doctrine contained in this work, will see that neither of these terms can be taken as ultimate. He will see that though the relation of subject and object renders necessary to us these antithetical conceptions of Spirit and Matter; the one is no less than the other to be regarded as but a sign of the Unknown Reality which underlies both.